Practical Embedded Security

Practical Embedded Security

Building Secure Resource-Constrained Systems

By

Timothy Stapko

ELSEVIER

AMSTERDAM • BOSTON • HEIDELBERG • LONDON
NEW YORK • OXFORD • PARIS • SAN DIEGO
SAN FRANCISCO • SINGAPORE • SYDNEY • TOKYO

Newnes is an imprint of Elsevier

Newnes

Newnes is an imprint of Elsevier

30 Corporate Drive, Suite 400, Burlington, MA 01803, USA

Linacre House, Jordan Hill, Oxford OX2 8DP, UK

Recognizing the importance of preserving what has been written, Elsevier prints its books on acid-free paper whenever possible.

Library of Congress Cataloging-in-Publication Data
Application submitted

British Library Cataloguing-in-Publication Data
A catalogue record for this book is available from the British Library.

ISBN: 978-0-7506-8215-2

For information on all Newnes publications
visit our Web site at www.books.elsevier.com

07 08 09 10   10 9 8 7 6 5 4 3 2 1

Printed in the United States of America

Contents

Preface

Living in a Connected World

1:37 AM. Hoover Dam, straddling the border of Nevada and Arizona, is quietly generating electricity for millions of Americans. The power plant, having recently been retrofitted with a new, remotely controlled automation system, is devoid of life, except for the blinking lights of the network hubs and automated hardware. Suddenly, the control room is ablaze with light, and the whirring of machines breaks the silence. The enormous floodgates open, a torrent of water rushing forth, sending a wave of destruction toward the unsuspecting communities downstream on the Colorado River. The turbines grind to a halt, plunging the desert into darkness. All the while, a teenager in Florida is laughing in the glow of his computer monitor.

Obviously, no one in his or her right mind would trust the control of Hoover Dam to a system with such gaping vulnerabilities, but the hyperbole of the example above does bring up an important point: as more and more of the world goes "online," we are putting more and more trust in the embedded systems that are designed to help us out. Obviously, something like the Hoover Dam would not be automated and connected to the Internet without a large investment in security, if it was automated at all. However, something far simpler, such as a home automation system, would likely not be subject to the same rigorous treatment as a vital hydroelectric power plant. This split between the security requirements of different embedded systems helps to illustrate the challenge of security design for embedded systems engineers. While the cutting edge of security is continually being pushed, low-end hardware and inexpensive systems are often left behind. However, these inexpensive systems are increasingly being networked and used to control more and more vital systems. This leads to an interesting and disturbing problem: Security implementations are often jealously guarded proprietary solutions that sell for thousands of dollars, which is directly at odds with the idea of using inexpensive microcontrollers. There are some options, such as various open-source implementations, but these can be unwieldy and are designed for PCs. If you want to design an inexpensive system and make it secure, there just are not many options.

One of the biggest problems with security in both the Hoover Dam example and home automation is the continual need for updates to keep up with malicious hackers. Anyone with a PC running Microsoft Windows knows about this from the continual stream of updates and patches for various security issues. One way to alleviate the continual update problem is to design security into the system and develop a solid application to begin with. The primary goal of this book is to introduce the users of inexpensive microcontrollers and embedded processors to the basic practical application of security and to provide some tools and pointers to assist in designing more secure applications with limited resources.

Many of the topics discussed in this book are covered in depth in hundreds of academic papers and tomes filled with arcane symbols. If you are interested in the mathematical underpinnings of cryptography, you are going to want to look elsewhere. However, if you work with microcontrollers or inexpensive embedded systems and security has been some-thing of interest but you have been intimidated by it, then this book is for you. Security is a hard problem, and a lot of very smart people have spent a lot of time working on it. The result is that the topic of security has taken on an intimidating air, especially when it comes to cryptography. This book aims to leverage the large body of work already done on security and adapt it for systems that usually aren't deemed powerful enough. As you will see, it is possible to implement security for some of even the most modest of architectures, such as porting AES to a PIC and using SSL on an 8-bit microprocessor (both of these are covered in extensive case studies of working implementations).

This book covers the practical side of implementing security for embedded systems, using publicly available and inexpensive proprietary implementations whenever possible. However, just having a cryptographic algorithm does not mean you have security. There are a number of issues to consider when using cryptography. We will cover some of them and hopefully provide some insight into how you can find them on your own.

Security in Shades of Gray

There is no such thing as perfect security. Think about it. As long as there is information that can be exploited, there will be someone trying to get it, and with enough resources, any security can be broken. The only way to be sure that information is completely "safe" is to destroy it and kill all the people who know about it, which obviously does no one any good. For this reason, secure systems are built using the idea of hazard tolerance—that is, each system has to have security that meets the requirements for the system. For example, a credit card system needs more security than your personal email (which we all know is completely *in*secure).[1] Unfortunately, security is an inherently difficult problem to solve,

---

[1] This may not be completely true anymore, as there are various applications (such as PGP) that provide encryption services for email. Generally speaking, though, you really should never send your credit card number via email.

since the worst problems are necessarily those that you cannot possibly predict. The only way to ensure a high level of security is to make your system as robust as possible, and keep it simple enough to understand so you can at least predict some of the more difficult problems. The fewer legitimate access points into your system you implement, the higher the probability it is safe. The more features and possible outside connections available, the more likely it is that you will have an unintended back door. Legitimate entry points need to be secured using a number of different mechanisms, depending on the desired level of security and the application. These mechanisms range from simple password schemes that provide only a small illusion of security to full security protocols that require vast amounts of computing power. Many of the mechanisms used to protect data in full security protocols, such as cryptography, usually require rather high levels of computing power, since they are based on powerful mathematical algorithms that require millions of calculations to be performed. Most security protocols work under the assumption that only the most powerful hardware is available. The problem with this assumption, of course, is that the most powerful hardware is very often, for economic or other reasons, not available.

Enter the *resource constrained system*. Embedded systems that utilize low-cost components may not have the resources required to implement one of these "true" security solutions. Many security protocols, algorithms, and mechanisms are built for the latest and greatest hardware—usually PCs or the most expensive embedded controllers and processors. The vendors will tell you that you need all that power to build a secure system, but the reality is that it really only depends on your application. You may not have access to that kind of power but still need security, so are you simply out of luck? This is why we introduced the chapter with the discussion on hazard tolerance: To build secure systems in a resource-constrained environment, we need to adapt the security or the application so that they work together without bringing the entire system to a halt (potentially a dangerous problem if the device is controlling a large automated punch press, for example). The idea behind this book is that it *is* possible to build secure *and* cost-effective systems.

Who This Book Is For

This book is for anyone interested in making the world more secure. Embedded systems make up the lion's share of the technology market (in volume, not necessarily in revenue) and are as pervasive as the products they help build and control. This book is particularly suited to embedded systems designers and engineers, but it may serve engineering managers as an introduction to a very important subject. Software engineers (embedded or not) are not the target audience, since many of the topics contained herein are covered in more depth in computer science courses and reference materials, but the case studies may still be of some interest. This book takes a practical approach to implementing security using available implementations and does not delve deeply into the mathematical and theoretical foundations of security. For that, the reader is encouraged to attend university computer security courses or read anything by Bruce Schneier.[2]

The content, though technical, should not be outside the reach of nonengineers (although knowledge of programming and Internet technologies will definitely help). The idea is to juxtapose technical content with a higher-level discussion of the subject to get the reader interested in the material and to think about the implications of deploying unsecured applications.

What This Book Is and What It Is Not

The goal of this book is to be a resource for all embedded systems designers—the first place to turn to when looking into security. The scope of computer security is so broad that no single text will provide all the answers you might need. This book aims to give the reader a start in the right direction, looking at some of the technologies available, providing a context in which to discuss those technologies, and giving a starting point for designing secure embedded systems. It should be considered as a first read for embedded security research, to get a quick overview of the challenges involved, and to get some ideas to move forward with. This book is organized so that the reader can quickly locate the information he or she needs and use it as a basis for the research into the project at hand.

This book is not a complete treatment of computer security. Indeed, it is not even a complete treatment of secure embedded systems design. The assumption is that the material presented will get the reader thinking about his or her own project, provide a platform of information to start off with. We will leave the goriest details to texts devoted to rigorous mathematical treatments and detailed security protocols that already flood the market (as an example, the de facto standard reference text for the Secure Sockets Layer by itself is over 400 pages long!).[3]

Why Embedded Security?

Some people may ask why this book is necessary. There are so many general-purpose texts on computer security that it seems one could simply buy a few books and be able to design a secure embedded system. The problem is, as has been mentioned previously, that these texts usually cover security under ideal conditions—that is, the hardware can support the mechanisms that are used and generally can support many mechanisms simultaneously. There are also some who believe that if you need security for a system, you should just buy

---

[2] Bruce Schneier is widely known as a foremost expert in computer security and is the author of several excellent books on the subject, most notably *Applied Cryptography*.

[3] *SSL and TLS: Designing and Building Secure Systems*, by Eric Rescorla.

the latest and greatest hardware and use standard (usually PC-centric) security protocols. The real world of embedded design does not always work like that. Economics plays a big role. Some applications require thousands or millions of units, and saving a few dollars on components really adds up. Why should we have to upgrade to expensive hardware just so that we can use a cookie-cutter implementation designed to work on a PC? In fact, many vendors will claim that you need the most feature-packed hardware package they offer or you will not have any security at all. Since security is considered "voodoo" or "black magic" by many people, these claims seem reasonable. This couldn't be farther from the truth. There is absolutely no reason we should not expect some level of security from even the most modest hardware. There are things that can be done to provide protection on any system, and we shouldn't have to choose between cost-effectiveness and peace of mind.

The key point to remember is that embedded security is *application dependent*. In the personal computer–dominated Internet, security solutions are typically designed to be general and flexible, reflecting the properties of the systems being protected (the general-purpose and flexible PCs). In the embedded world, the systems needing protection have unique properties and are particularly suited to specific applications. The security needed by these applications is similarly unique and specific. The general solutions typically employed in the existing Internet often do not work "out of the box" for embedded systems, and simply porting existing security protocols can lead to code bloat for features that are unnecessary. Instead, the security for an embedded system needs to be specifically tailored to the particular application being developed. This is not to say that existing protocols and policies cannot be used (and indeed *should* be used), but rather that we need to adapt Internet security to an embedded world through analysis of the applications to which these concepts are applied.

The reader is encouraged to use this book as a way to start learning how to look at security for embedded systems, rather than as a universal solution to all your security needs. Reading the book cover to cover will definitely increase your knowledge of the subject matter, but it is by no means necessary in order to get what you need out of the book. It is recommend that you read the following three chapters, covering computer security fundamentals, Internet security, and the principles of embedded Internet security, respectively. These chapters will give you a foundation for the rest of the text. If you are already familiar with computer and Internet security, then go ahead and skip to Chapter 3 (where we cover security algorithms and protocols) unless you need a refresher (which is recommended anyway). After reading those chapters, the rest of the book is organized so that you can easily find a particular topic of interest, or just continue reading through all the chapters to get a comprehensive coverage of the content. We will look at the various aspects of embedded Internet security, from communications, to software security implementations, to hardware for security.

Roadmap

Chapter 1 introduces (or reintroduces) readers to the basics of computer security, with some light theoretical background, a look at the different subfields within computer security, and, most importantly, a look at the security mechanisms that will be covered in the rest of the book. This information is provided as background for readers unfamiliar with security and cryptography. Those readers with some background may wish to skip ahead to later chapters.

Our security introduction will begin with some light theory to give the reader a foundation for the rest of the chapters. The basic premise of computer security is *access control*—who can and cannot control and access some particular application or data. We will look at the theory behind access control and introduce some analysis techniques based on the access control matrix model.

In Chapter 2, we will look at some low-level networking protocols and what to look out for in using them. Part of the reason for looking at this low-level functionality is to get used to analyzing the protocols you intend to use. Without some understanding of how the low-level things work, you cannot have much assurance that your application is secure.

Some of the topics to cover are TCP/IP. UDP, Serial communications, PPP, and Ethernet (wireless protocols will be discussed later). There will be a brief description of each protocol, what it is used for, and how it works with security mechanisms. There are definite differences between all of these networking technologies that may lead to issues when attempting to secure an application. Choosing a communications technology wisely can save headache and extra work later. We will try to sort out the mess of available options in an attempt to help the reader choose the technology that is best for them.

In Chapter 3 we will look at the world of Internet security, from simple hashing techniques used in web-based applications to full-blown security protocols like SSL. In this chapter we look at various standard cryptographic algorithms and how they are used. This chapter covers the algorithms that are the building blocks of secure networked systems.

Chapter 4 is all about SSL. The Secure Sockets Layer is so important to Internet security that we devote an entire chapter to it. SSL is the de facto standard for secure Internet transactions. It has achieved this status by being not only secure, but being highly generic as well. SSL exists in the network layer between TCP and your application, providing blanket security to all data transferred over the network. The API for SSL is typically very similar to the standard network sockets API (POSIX-style). For this reason, it is simple to transform any plain TCP/IP application into a secure Internet application with very little effort. We will look at how to implement SSL for embedded platforms and cover the options we have in adapting the protocol to work for specific applications. SSL is very component-driven, and we can use this to our advantage.

Chapter 5 covers security from the embedded systems developer's point of view. First and foremost, the primary concept that will be introduced here and reinforced throughout the text is that the embedded systems we are covering are *NOT PCs or expensive devices running Windows or Linux*. In this chapter we will also introduce strategies for choosing algorithms and protocols for a particular platform and for choosing a platform based on security requirements. There are many tradeoffs to be considered, and we will look at some examples to get an idea of what is available to the embedded developer. We will look at some of the features, pros and cons, of different protocols and algorithms. The reader can then use this section as a guide to using the rest of the book for their applications.

Also in Chapter 5, we will look at embedded security protocols—or, rather, the lack of them. The problem is that unlike a PC, which can support a myriad of protocols and algorithms simultaneously, each embedded device can only have one or two protocols resident at any given time. We will spend the rest of this chapter looking at why designing an embedded security protocol is extremely difficult. The requirements vary greatly between applications, and the capabilities of the hardware have a similar variance. This section will provide a justification for the treatment of the protocols and algorithms throughout the remainder of the text.

By Chapter 6 the reader should have a fairly decent understanding of the concepts involved in securing a system, and he/she should have a relatively clear picture of why securing embedded systems represents a significant challenge. This chapter covers networking technologies that will allow a device to be connected wirelessly to the Internet or each other. The vast majority of new resource-constrained networked systems will likely be in the wireless and cellular arena. Wireless networking has taken off as a hot new technology and represents one of the fastest-growing markets in the industry. We will look at protocols that are commonly used and appropriate for embedded development, such as 802.11 (Wi-Fi), Bluetooth, and ZigBee. Beyond this, we will look at each protocol for the purposes of securing data being sent and received.

In Chapter 7 we look at client/server applications and their relevance to the embedded world. The World Wide Web is by far the most recognizable and widely used application of the Internet and represents the fundamental client/server relationship. As such, it stands to reason that a large number of networked embedded applications will have some type of Web service functionality. For this reason, we will start our tour of secure embedded applications with a look at the Web and other client/server applications.

Chapter 8 covers some basic ideas and common pitfalls related to the optimization of security mechanisms. Here we look at cryptographic algorithms and how to make them work for embedded machines. Cryptography presents the biggest challenge to any security solution to be implemented for a resource-constrained embedded device because of the requirements of these computationally complex algorithms. Cryptography is notoriously expensive when it comes to clock cycles, and many algorithms are not too friendly to small storage devices (code and data). We will look at various strategies for choosing algorithms for specific applications and look at some specific algorithms, as well as some strategies to avoid some of the more problematic resource issues.

In Chapter 9 we look at hardware alternatives that allow embedded developers to get the most out of their systems. These alternatives include hardware assistance and complete hardware security solutions. Hardware assistance involves dedicated circuitry that provides part or all of the functionality of a security mechanism at the speed of hardware. Complete hardware solutions include entire security protocols implemented in silicon for the maximum possible performance. Both of these ideas can be used to augment an embedded system without sacrificing the affordability of the base hardware. We will also briefly cover physical security. When applying security to an embedded system, this is an important point that may be overlooked. The reason for this is that, unlike PCs and servers, which can be deployed in controlled environments, embedded devices are deployed practically every-where. We will look at some different technologies to consider when choosing an embedded device for a secure embedded system.

In Chapter 10 we cover some miscellaneous issues with security, such as programming gotchas in languages like C and recognizing and dealing with attacks. We finish up the chapter with a brief mention about the political issues regarding cryptography and exported applications. We start the chapter by looking at how development tools and languages factor into the security of an application. Some languages, such as Java, provide some built-in safeguards that help with overall security, such as strict typing. C is the predominant language used for many embedded devices, but it suffers from many shortcomings that make it difficult to use in implementing a secure system. We will look at some of the features of different languages and discuss what to look for when choosing tools and languages and when developing software for your system. We will also briefly cover attacks and how to deal with them. Any secure system, or for that matter, *any* system connected to a network (the Internet or proprietary) *will* be subject to attacks—both intentional (i.e., malicious hackers) and inadvertently (i.e., heavy network traffic leading to Denial of Service; we will classify accidents as attacks to simplify this chapter). We will look at how to deal with future attacks, currently occurring attacks, and the aftermath after a successful attack. Finally, we will briefly look at export issues, primarily to inform the reader of their existence, since dealing with political issues is far beyond the scope of this text. We close out Chapter 10 with some further reading and a discussion about the future of embedded security.

In Chapters 11 and 12 we look at some application case studies. In these chapters we will develop two applications from requirements to implementation that use inexpensive hardware but also need some level of security. These chapters are provided to help the reader see embedded security applied to real-world implementations. We will go through the entire development of each application, from initial requirements, to developing a security policy, verifying the design, and finally deploying the application. There is complete code for the two applications, with full listings of the programs (not including library code) in Appendix A. The applications are working implementations on real hardware (using a PIC and the 8-bit Rabbit 4000 microprocessor) that provide some insights into the development of a secure application for a platform with limited resources.

CHAPTER 1

Computer Security Introduction and Review

This chapter is intended to provide a quick introduction to computer security for embedded systems engineers who may not have a formal background in computer science and computer security. For the more advanced reader, this chapter serves as a review of computer security before delving into the later material. This chapter is by no means a complete treatment of the theory behind computer security—literally hundreds of books have been written on the subject—but it should at least provide a basic context for all readers. At the end of the chapter, we will provide a list of further reading for readers wanting a deeper treatment of the theory. We will briefly touch on the most important concepts, spending most of the discussion on those ideas that are most pertinent to embedded and resource-constrained systems.

Computer security is a rapidly evolving field; every new technology is a target for hackers, crackers, spyware, trojans, worms, and malicious viruses. However, the threat of computer attacks dates back to the earliest days of mainframes used in the 1960s. As more and more companies turned to computer technology for important tasks, attacks on computer systems became more and more of a worry. In the early days of the Personal Computer, the worry was viruses. With the advent of the World Wide Web and the exponential expansion of the Internet in the late 1990s, the worry became hackers and denial of service attacks. Now, at the dawn of the new millennium, the worry has become spam, malware/spyware, email worms, and identity theft. All of this begs the question: How do we protect ourselves from this perpetual onslaught of ever-adapting attacks?

The answer, as you may have guessed, is to be vigilant, staying one step ahead of those who would maliciously compromise the security of your system. Utilizing cryptography, access control policies, security protocols, software engineering best practices, and good old common sense, we can improve the security of any system. As is stated by Matt Bishop,[1] computer security is both a science *and* an art. In this chapter, we will introduce this idea to embedded systems engineers and review the basic foundations of computer security to provide a foundation for the rest of the book.

---

[1] Author of *Computer Security: Art and Science.*

What Is Security?

To begin, we need to define *security* in a fashion appropriate for our discussion. For our purposes, we will define computer security *as follows*:

> *Definition:* Computer Security. *Computer security is the protection of personal or confidential information and/or computer resources from individuals or organizations that would willfully destroy or use said information for malicious purposes.*

Another important point often overlooked in computer security is that the security does not need to be limited to simply the protection of resources from malicious sources—it could actually involve protection from the application itself. This is a topic usually covered in software engineering, but the concepts used there are very similar to the methods used to make an application secure. Building a secure computer system also involves designing a robust application that can deal with internal failures; no level of security is useful if the system crashes and is rendered unusable. A truly secure system is not only safe from external forces, but from internal problems as well. The most important point is to remember that *any* flaw in a system can be exploited for malicious purposes.

If you are not familiar with computer security, you are probably thinking, "What does 'protection' actually mean for a computer system?" It turns out that there are many factors that need to be considered, since any flaw in the system represents a potential vulnerability. In software, there can be buffer overflows, which potentially allow access to protected resources within the system. Unintended side effects and poorly understood features can also be gaping holes just asking for someone to break in. Use of cryptography does not guarantee a secure system either; using the strongest cryptography available does not help if someone can simply hack into your machine and steal that data directly from the source. Physical security also needs to be considered. Can a malicious individual gain access to an otherwise protected system by compromising the physical components of the system (this is especially important for embedded systems)? Finally, there is the human factor. Social engineering, essentially the profession practiced by con artists, turns out to be a major factor in many computer system security breaches. This book will cover all of the above issues, except the human factor. There is little that can be done to secure human activities, and it is a subject best left to lawyers and politicians.

What Can We Do?

In the face of all these adversities, what can we do to make the system less vulnerable? Next we will look at the basics of computer security from a general level to familiarize the reader with the concepts that will be reiterated throughout the book. Even the experienced reader may find this useful as a review before we delve into the specifics of network and Internet security in Chapter 2.

Access Control and the Origins of Computer Security Theory

In their seminal computer security paper, "The Protection of Information and Computer Systems," (Saltzer 1976) Saltzer and Schroeder recorded the beginning concepts of access control, using the theory that it is better to deny access to all resources by default and instead explicitly allow access to those resources, rather than attempt to explicitly deny access rights.[2] The reason for this, which may be obvious to you, is that it is impossible to know *all* the possible entities that will attempt access to the protected resources, and the methods through which they gain this access. The problem is that it only takes one forgotten rule of denial to compromise the security of the entire system. Strict denial to all resources guarantees that only those individuals or organizations given explicit access to the resources will be able to have access. The system is then designed so that access to specific resources can be granted to specific entities. This control of resources is the fundamental idea behind computer security, and is commonly referred to as *access control*.

Over the years, computer scientists have formalized the idea of access control, building models and mathematically proving different policies. The most versatile and widely used model is called the *access control matrix*. Shown in Figure 1, the access control matrix is comprised of a grid, with resources on one axis and entities that can access those resources on the other. The entries in the grid represent the rights those entities have over the corresponding resources. Using this model, we can represent all security situations for any system. Unfortunately, the sheer number of possibilities makes it very difficult to use for any practical purposes in its complete form (representing all resources and possible users). We can, however, simplify the concept to represent larger ideas, simplifying the matrix for looking at systems in a consistent and logical manner. This is a concept that can be applied throughout the rest of the book to represent the resources and users that will be acting on

	Alice (manager)	Bob (IT admin)	Carl (Normal user)	Donna (Temporary user)
C:\Users\Alice	RWG	RWG		
C:\Users\Bob	R	RWG		
C:\Users\Carl	R	RWG	RWG	
C:\Users\Donna	R	RWG		RW

R = Can read from directory
W = Can Write to directory
G = Can grant other users read, write, or grant permission

Figure 1: Access Control Matrix

---

[2] This idea, by the authors' admission, had been around since at least 1965.

the systems we are looking to secure. Having a logical and consistent representation allows us to compare and contrast different security mechanisms and policies as they apply to a given system.

In order to understand what an access control matrix can do for us, we will define a few rights that can be applied to users and resources. For our purposes, we will not give the access control matrix a complete formal treatment. We will instead focus on the rights and concepts that are directly applicable to the systems that we are looking at. For a more complete treatment of the theory behind access control matrices, see *Computer Security: Art and Science* by Matt Bishop.

The rights we are most interested in are *read, write,* and *grant.* These rights are defined as follows:

Read—The ability to access a particular resource to gain its current state, without any ability to change that state.

Write—The ability to change the state of a particular resource.

Grant—The ability of a user to give access rights (including grant privileges) to another user.

The rights defined here are a simplification of the full model, but will serve to help explain different security policies. The most important of these rights is *grant*. This right allows an expansion of rights to other users, and represents a possible security problem.

Given the rights defined as part of the access control matrix model, we can now analyze any given system and how secure it is, or is not. Using the matrix built from our system, we can mathematically guarantee certain states will or will not be entered. If we can prove that the only states the system enters are secure (that is, no unauthorized entities can get rights they are not entitled to, purposefully or inadvertently), then we can be sure that the system is secure. The problem with the access control matrix model, however, is that it has been proven this problem is undecidable in the general case when any user is given the *grant* right, since it opens the possibility of an inadvertent granting of a right to an unauthorized user. This does not mean the model does not have its uses. We will study different mechanisms and policies using this model because it simply and efficiently represents security concepts. In the next section, we are going to look at security policies and how they are designed and enforced in common applications.

Security Policies

The access control matrix provides a theoretical foundation for defining what security is, but what it does not do is provide a practical method for implementing security for a system.

For that, we need a security policy. The idea behind a security policy is simple: It is a set of rules that must be applied to and enforced by the system to guarantee some predefined level of security. Analogous to a legal system's penal code, a security policy defines what entities (people and other systems) should and should not do. When designing an application that you know will need security, part of the requirements should be a list of all the things in the system that need to be protected. This list should form the basis for the security policy, which should be an integral part of the design.

Each feature of the application must be accounted for in the policy, or there will be no security; as an example, think of the networked home thermostat example. If the security policy covers only the obvious features that may be threatened, such as a web interface, it might miss something more subtle, like someone opening the thermostat box and accessing the system directly. If the thermostat has the greatest Internet security in the world, but it sits wide open for someone to tinker with if he or she is physically there, it is probably in violation of the intended security policy, not the actual one. In this example, the security policy should include a rule, which can be as simple as a single statement that says the physical interface should have the same password scheme as the network interface. To take it a step further, the policy might also include a rule that the thermostat box should be physically locked and only certain personnel have access to the key.

Though it seems like there should be, there are no rules governing security policies in general. Certain organizations, such as corporations and the government, have certain guidelines and certifications that must be followed before an application is considered "secure," but even within a single organization, the security policies will likely differ. The problem is, as we will repeat throughout this book, that security is application-dependent. Although this means there is no official template to start from, a good starting place for a security policy would be the initial requirements document for the application. Each feature should be analyzed for its impact to the system should it ever be compromised. Other things to take into account are the hardware used, the development tools and language used, the physical properties of the application (where is it), and who is going to be using it.

When developing your security policy, think of your application as a castle. You need to protect the inhabitants from incoming attacks and make sure they are content (at least if they are not happy living in a castle); see Figure 2. Your policy is then a checklist of all the things that might allow the inhabitants to come to harm. There are the obvious things (get them out of the way first) like locking the castle gate and requiring proof of identity before allowing someone to enter (the password in an electronic application), or making sure the inhabitants can do their jobs (the application performs its tasks). However, to truly develop a useful policy, you have to think a little like the enemy. What are some other possibilities? Well, the enemy might not care about taking total control of the castle, and instead prevent it from functioning properly by stopping incoming traders from reaching the castle (denial

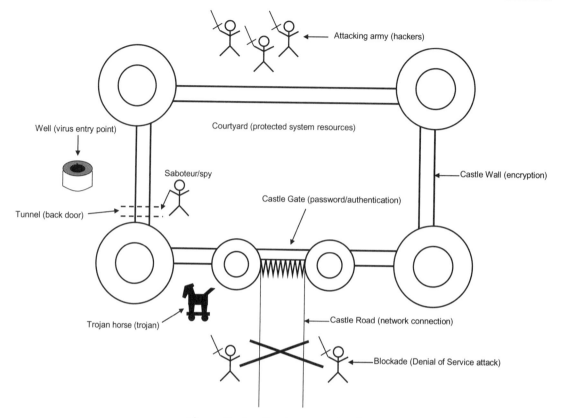

Figure 2: Application as a Castle

of service attack). A more subtle attack might be something that is not noticed at first, but later becomes a serious problem, like hiring an inhabitant to sabotage the defenses (disgruntled employee making it easier for hackers), or poisoning the water supply (viruses). The enemy may also rely on cleverness, such as the mythical Trojan horse (Trojan horses, basically viruses that give hackers a doorway directly into a system). The enemy may not even be recognizable as an enemy, in the case of refugees from a neighboring war-ravaged country suddenly showing up and eating up all the available food and resources (worms like Blaster and Sasser come to mind). The possibilities go on and on, and it can be daunting to think of everything that might be a problem.

In the castle example above, we paralleled some physical security problems with related computer security problems. This illustration may be useful to think of when developing a security policy, since many problems in security are very similar in vastly different contexts. It also helps to know what the most common attacks are, and weigh the likelihood of an attack versus the damage that the attack will wreak on your application should the attack be successful. One thing to try would be to think of the worst things your system could do

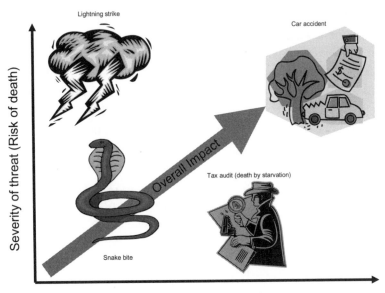

Figure 3: Severity of Threats

(thermostat cranked up to 120°, opening the floodgates on the dam at the wrong time, etc.), and rank their severity on a scale (1 to 10, 1 to 100, etc.). Then think like a hacker and come up with attacks that each result in one or more of those scenarios, and assign a probability based upon the relative difficulty of the attack (this is very hard to do, see below). Take the probability of each attack that results in each of those scenarios and multiply it with the severity level. This will give you an idea of the relative importance of protecting against a particular attack. The problem with this system is that the probability of an attack is very hard to determine (close to impossible for any sufficiently complex system). There is almost always someone smarter that can do something easily that you think is difficult (this is where feedback from colleagues is important). Furthermore, the probabilities you assign might change if another less "important" attack is successful. Given this, it still can be useful to assign relative weights to the things that can happen, and focus on those that would result in the most damage. In Figure 3, we see the relative impact of different threats in the real world, as compared to the probability of the threat occurring.

As can be inferred from the discussion above, producing a security policy is a creative process, much like painting a picture or writing music. Like other creative processes, it is therefore beneficial to recruit critics who will tell you what is wrong. In the case of security, the critic would likely be an expert security consultant who you hire to look over your application design (this is recommended for applications where security is truly important). If you think about it for a bit, it should be pretty easy to see that the more people that look at your policy, the better chance you have of catching problems. It is also a time to be paranoid. If a consultant wants to keep everything secret and not reveal what he or she is doing and then fire everybody, you should be able to see what is going on (make the consultant explain everything to you). In any case, more eyes will result in better security, even though this seems counterintuitive. Keeping things secret (security through obscurity) would seem to provide an additional level of security, and indeed it does, but it detracts from the overall security of the system. The reason for this is that it is unlikely that an individual (or even an organization) can think of all the possibilities for security breaches alone, it requires different viewpoints and experience to see what might lead to problems later.[3] Over and over, it has been proven that hiding the workings of security mechanisms does not work. Almost all of today's most widely used algorithms and protocols are wide open for review by all. A good security policy should be able to stand on its own without having to be kept secret—just don't share your passwords with anyone!

Throughout the rest of the book, keep the idea of implementing security policies in mind, since what we are really talking about is the enforcement of the rules put forth in a security policy. The enforcement of a rule may include cryptography, smart engineering, or basic common sense. This book is about the tools and mechanisms that can be employed for the enforcement of security policies. In the next section we will look at one of the most important tools in our security enforcement arsenal: cryptography.

Cryptography

In the last section, we looked at security policies and how they define the access that users have to resources. However, we did not look at the mechanisms that are used to enforce these policies. In this section, we introduce and describe the most important of these mechanisms: cryptography. *Cryptography* is the science of encoding data such that a person or machine cannot easily (or feasibly) derive the encoded information without the knowledge of some secret *key*, usually a large, difficult to calculate number. There are several forms of cryptography, some new, some dating back thousands of years. There is proof that ancient civilizations (namely, Egypt and Rome) used primitive forms of cryptography to communicate sensitive military information without the fear that the courier might be captured by the enemy. In modern history, the most famous form of encryption is also one of the most infamous—the Enigma machines used by Germany in World War II that were broken by the Allies. An example Enigma machine is shown in Figure 4. The breakdown in security allowed the US and Britain to gain important intelligence from the Germans, leading

---

[3] One possible exception to this is the US National Security Agency (NSA), since they employ a large number of security experts, giving them a definite advantage over other organizations. If you are the NSA, then feel free to keep everything secret; otherwise you may want to recruit some help.

Figure 4: Enigma Machine (from www.nsa.gov)

directly to defeat of Germany. The Enigma example also illustrates some important concepts about cryptography. Several things had to happen to break the system, such as the Allies obtaining an actual Enigma machine and German soldiers not following the security policy required by the system (essentially always using the same password). Those breaches are hard to guard against, so the system must be designed to minimize the impact of those attacks. The Enigma system was fundamentally flawed because these relatively minor breaches[4] led to complete failure of the system—as a result, the Allies won the war.

The newest form of encryption is *quantum cryptography*, a form of cryptography that utilizes the properties of subatomic particles and quantum mechanics to encode data in a theoretically unbreakable way. There are some actual practical applications of quantum cryptography in existence today, but they are severely limited and far too expensive for all but the most important applications. For this reason, we only mention it here in this chapter and again in the conclusion, and will not include it in our embedded systems discussions.

---

[4] Not to belittle the efforts of the allies to obtain an Enigma machine, but how do you protect a machine like that—one careless move and the enemy has one. In the Internet age, it is even easier to obtain a working "machine" for any number of security systems, since source code is easy to copy.

Symmetric Cryptography

To start our discussion of cryptography, we will start with the oldest and most prevalent form of encryption: *symmetric-key cryptography*. This is the type of cryptography practiced by ancient civilizations and was the only true type of cryptography until the last century. Symmetric-key cryptography is characterized by the use of a single secret *key* to encrypt and decrypt secret information. This use of a single key is where the name *symmetric* came from, the same algorithm and key are used in both directions—hence the entire operation is symmetric (we will see the opposite of symmetric cryptography, called *asymmetric cryptography,* in the next section).

To illustrate what symmetric cryptography is, we will use the classic computer security characters Alice, Bob, and Eve. Alice wishes to send a message to Bob, but Eve could benefit from having the information contained in the message, to Alice's detriment. To protect the message from Eve, Alice wants to employ symmetric-key cryptography. However, since the same key needs to be used for both encryption and decryption, Bob needs to have a copy of the key so he can read the message. This works fine *if* Alice and Bob met earlier to exchange copies of the keys they want to use. It would also work if they had a reliable and trustworthy courier to deliver the key. However, if Alice attempted to simply send a copy of her key to Bob (using a questionably trustworthy method, such as email), it is very likely that Eve would be able to gain a copy of the key while in transit. To see symmetric-key cryptography in action, see Figure 5.

Obviously, symmetric-key cryptography has some serious drawbacks for computer security. For instance, how do you give a secret key to someone you have never met (which is exactly what needs to happen for e-commerce)? Also, what do you do if your key is

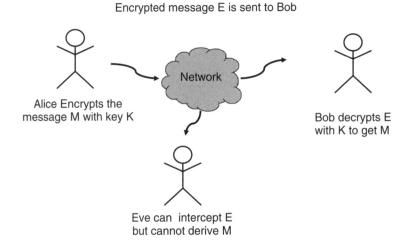

Figure 5: Symmetric-key Encryption Example

compromised or stolen? How do you get a new key to the recipient of your messages? Despite these drawbacks, however, symmetric-key cryptography *does* have a place in computer security. As it turns out, symmetric-key algorithms are the simplest, fastest cryptographic algorithms we know of. In a world built on bandwidth, this speed is a necessity. Symmetric-key algorithms also really make a difference in the embedded world. Since some algorithms can be implemented in a few lines of C, and can be optimized to be incredibly fast (without compromising the integrity of the security—more on this in later chapters), symmetric-key cryptography finds a welcome home in the embedded world. Later on we will cover these algorithms in more detail and show how they can be integrated into an embedded system—both as a stand-alone feature and as part of a larger security solution.

Public-Key Cryptography

In the last section, we covered the oldest and most common form of cryptography. As you may have been wondering, if symmetric-key cryptography is symmetric, is there an asymmetric opposite? The answer, as you may have also guessed, is yes. *Asymmetric cryptography* is the conceptual opposite of symmetric-key cryptography. Asymmetric cryptography is usually referred to by its more common name (which is also more descriptive), *Public-key cryptography*.

Public-key cryptography is a very new method for encrypting data, developed in the 1970s as a response to the limitations of symmetric-key cryptography in an online world (recall that the Internet was in its infancy at that time). The basic idea behind the creation of public-key cryptography was to create a "puzzle" that would be very difficult to solve unless you know the trick (the key), in which case solving the puzzle would be trivial—solving the puzzle reveals the message. Using the analogy of a jigsaw puzzle, imagine that you create a blank puzzle and send the completed puzzle to your friend. Your friend writes her message on the completed puzzle, then takes the puzzle apart and returns the pieces to you (encoded with the message). If the puzzle has some trick needed to solve it, and only you know that trick, you can be assured that your friend's message arrives at your doorstep unspoiled by eavesdropping. You also have the added benefit of knowing if the message was tampered with, since you will not be able to solve the puzzle successfully if any of the pieces change. We will now look at public-key cryptography as it is implemented today.

Public-key algorithms use different keys for both encryption and decryption (hence the asymmetry), and one of these keys is typically referred to as the *public-key*, since this key is usually published in some public place for anyone to access. This may at first seem counter-intuitive, why would you want to *publish* your keys? Wouldn't that mean anyone can access my secret data? As it turns out, you are only publishing *half* of your total key—the part used to encrypt data (recall the puzzle example—you publish the puzzle, but keep the trick to solving it to yourself). We will illustrate public-key cryptography using Alice, Bob, and Eve once again.

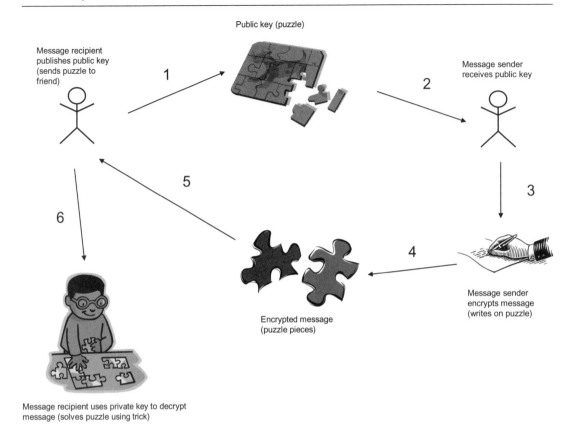

Figure 6: Jigsaw Puzzle Analogy for Public-Key Encryption

Alice wants to send another message to Bob, but they have not made arrangements to exchange a secret key in order to do symmetric-key cryptography to protect the message. Fortunately, both Alice and Bob have a public-key algorithm at their disposal. Alice can then request that Bob publish his public-key on a reliable website (the site *must* be reliable; we will look at this problem further when we discuss *Certificate Authorities* later on). Alice then downloads Bob's key and uses it to encrypt her message. After the encryption, no one can read the message but Bob. Not even Alice can decrypt her own message. The idea behind this is that Bob keeps half of his key secret (his *private* key), the part that can decrypt information encrypted using the public-key. Eve, who has been watching the transaction, has complete access to Bob's public-key, but is helpless if she tries to read Alice's message, since she does not have Bob's private key. Alice's message is therefore delivered safely to Bob. For an illustration of public-key cryptography in action, see Figure 7. Further on in the next section, we will revisit public-key cryptography and see how some algorithms can be used in a novel way to prove who sent a message when we look at authentication techniques.

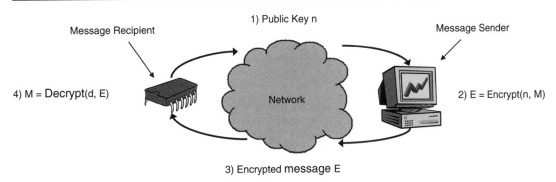

Figure 7: Public-Key Encryption

Data Integrity and Authentication

Cryptography is a useful tool in securing a system, but is just that, a tool, and there are many other tools available for security. To enhance cryptography in a practical system, and sometimes even replace it, we can use mechanisms and methods that protect data integrity and authenticate entities. Sometimes, one mechanism solves both of these problems. The first mechanism we will look at is what is called a *cryptographic hash* or *message digest algorithm*. Following that, we will discuss the inverse of public-key encryption; it turns out that reversing public-key algorithms (using the decryption key to encrypt a message, and using the encryption key to decrypt) allows us to do a form of authentication that is very useful in practice. We will also discuss methods and mechanisms of providing trust, such as digital signatures and certificates, and Public-Key Infrastructure (PKI).

Message Digests

At a very basic level, a message digest algorithm is simply a hash function, enter some arbitrary data of arbitrary size, and the hash algorithm spits out a fixed-size number that is relatively unique for the input given (note that we use the phrase *relatively unique*, as it is impossible to create a perfect hash for arbitrary input—a perfect hash can only be created if we can restrict the input to known values). What makes a hash function into a message digest is a level of guarantee that if two input datum are different (even by a single *bit*), then there is a predictably small possibility of a hash collision (those two messages generating the same hash). This property is quite important for message digests, and this will become apparent when we look at how message digests are used.

So it is great and wonderful that we have this algorithm that can take any size message and turn it into a much smaller number—so what? Remember that there is a *very small* probability of two messages generating the *same* hash value, and the chances of those two messages both containing legitimate data is even smaller. Using this knowledge, if we can provide a hash value for a message, then anyone can take the message, hash it on his or her machine, and verify that the locally generated hash and the provided hash match up. If they don't, then the message has been altered, either accidentally (some transmission error where data was lost or corrupted), or intentionally (by a malicious attacker). If used appropriately, the message digest mechanism gives us a fairly strong guarantee that a message has not been altered.

The guarantee given by hash algorithms is not perfect, however. Recently, there have been some advances in the mathematical analyses of both of the most commonly used algorithms, MD5 and SHA-1. For most intents and purposes, MD5 is considered insecure, and SHA-1 is not as secure as previously thought.[5] However, the attacks on these algorithms require a lot of computing power and dedicated attackers. This may be an issue for banking transactions or other information that has a long lifetime, but for many applications, the level of security still provided by these algorithms may be sufficient. The caveat here is that if you choose an algorithm now that is known to be flawed, in the near future it is likely that, with advances in mathematics and computer technology, the algorithm will be completely useless. The problem faced by everyone right now is that there are no readily available replacements for MD5 or SHA-1. Most applications today, however, were built with some safeguards in the anticipation of these types of compromises. For example, the Transport Layer Security protocol (TLS—the IETF[6] standard for SSL—the Secure Sockets Layer protocol, which we will discuss in depth in Chapter 4) uses an algorithm called HMAC, which wraps the hashing algorithm with some additional steps that allow the hashing algorithm to have some mathematically provable properties. We will discuss HMAC in more detail when we look at SSL in Chapter 4. This additional layer of security present in some current applications should help keep those applications fairly safe until suitable replacements are found for MD5 and SHA-1. The attacks against MD5 and SHA-1 also illustrate the benefits of allowing the public access to the workings of the algorithms, since these attacks likely would not have been found and exposed. We now know that these algorithms are flawed and can take additional precautions, whereas had the algorithms been proprietary, the attacks could have been discovered by an evildoer and we would never have known. It is even possible, albeit unlikely, that the faults were discovered earlier by the bad guys, but now we know what to look for, thus mitigating the potential effects of the attacks.

We have seen how hash algorithms are useful for protecting the integrity of data, but there is another common use for these handy algorithms, and that is *authentication*. Authentication is the ability to verify the correctness of some data with respect to a certain entity. For

---

[5] Several academic papers have shown various weaknesses in both MD5 and SHA-1. There is even source code that demonstrates a hash collision in MD5. For the most part, however, these attacks remain mostly academic.

[6] Internet Engineering Task Force—the organization that oversees many of the publicly available standards for networking and security related to the Internet (www.ietf.org).

Figure 8: Driver's License Security

example, when someone is "carded" when trying to purchase a beer, the bartender is attempting to authenticate the customer's claim that he or she is actually old enough to purchase alcohol. The authentication comes in the form of a driver's license. That license has certain guarantees associated with it that allow the bartender to authenticate the customer's claim. In this example, the authentication comes from the government-issued identification, which is difficult to counterfeit, and is backed by certain laws and organizations—see Figure 8. In the world of the Internet, it is not so simple, since we do not often see who it is that we are actually communicating with (and in many cases, the "who" is simply a server tucked in a back room somewhere!). A driver's license does you no good if you are Amazon.com and you want to prove who you are to an online shopper. This is where a novel use of public-key cryptography becomes a very valuable tool.

It turns out that if we reverse some public-key algorithms, using the private key to *encrypt* instead of decrypt, we can use a trusted version of the corresponding key to verify that a message came from the private key's owner (assuming, of course, that the owner has not lost control of that key). The reason for this is the same property that makes public-key cryptography possible in the first place. If we use a private key to encrypt some data, then only the associated public-key will be able to decrypt that data correctly. If the recipient of a message possesses a known good, trusted copy of the sender's public-key, and if the message decrypts correctly using that key, the recipient can be almost certain that the message came from person who provided the public-key in the first place. See Figure 9 for an example. Using this mechanism, we can provide a level of trust in network communications, as long as we have a secure way to gain access to trusted copies of public-keys. We will discuss current approaches to this problem further on, but first we will look at public-key authentication in practice.

Digital Signatures

As we mentioned before, public-key cryptography is horribly inefficient. For this reason, the authentication method we just looked at would not be practical if every message had to be encrypted using the public-key algorithm. In order to make public-key authentication practical, the *digital signature* was invented. A digital signature is simply a hash of the data to be sent (using one of the message digest algorithms) encrypted using the public-key

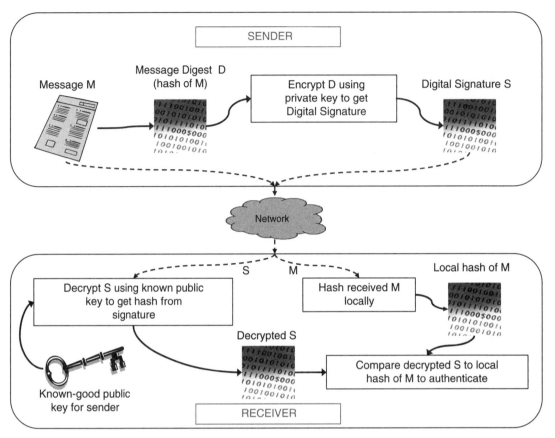

Figure 9: Public-Key Authentication

authentication method. By encrypting only the fixed-size message hash, we remove the inefficiency of the public-key algorithm and we can efficiently authenticate any arbitrary amount of data. Digital signatures are not fool-proof, however, since they rely on hashing algorithms that may have weaknesses, the public-key must be trusted, and the private key must always remain private. In practice, however, the digital signature does provide some level of security, and in fact, forms the basis of trust for most of the Internet and e-commerce. In the next section, we will look at how digital signatures are used in practice by many protocols and applications.

Digital Certificates

So now we have all sorts of nifty mechanisms for encrypting data, protecting the integrity of data, and authenticating the source of that data, but how do these all work together to provide security for our applications? The most common use of the digital signature for authentication is a part of a *digital certificate*. A digital certificate consists of three primary sections: information about the owner (such as real or company name, Internet address, and physical address), the owner's public-key, and the digital signature of that data (including the public-key) created using the owner's private key. Digital certificates are typically encoded using a language called ASN.1 (Abstract Syntax Notation), developed by the telecommunications industry in the late 1970s, and more specifically, a subset of that language called Distinguished Encoding Rules (DER). This language was designed to be flexible, allowing for any number of extensions to the certificate format, which are created and used by some users of digital certificates to provide additional information, such as which web browsers should be able to accept the certificate. The public-key is stored using a base-64 encoding.

A digital certificate is provided by sending an application at the start of a secure communication. The receiving application parses the certificate, decrypts the digital signature, hashes the information, and compares it to the decrypted signature hash. If the hashes do not match, then the certificate has been corrupted in transit or is a counterfeit. If the hashes do match, then the application does a second check against a field in the data section of the certificate called the *Common Name (CN)*. The common name represents the Internet address (url or IP address) of the sending application. If the common name and the address of the sending application do not match, then the receiving application can signal a warning to the user. Finally, a valid date range can be assigned to the certificate by the owner that will tell the recipient whether the certificate should be trusted at a certain time. If the current date (today) is between the initial date and the expiration date indicated in the certificate, then the certificate is considered valid (this does not supersede the other checks). Note that all these checks are up to the receiving application; the recipient of a digital certificate can choose to ignore all checks, or only perform one or two, but the guarantee of security is obviously not as strong if any failed checks are ignored. The digital certificate provides a useful mechanism for authentication, assuming that the public-key in the certificate can be trusted. This leaves us with a problem, however. If we only get the public-key as part of the digital certificate, how can we be certain that the person who sent the certificate is who they say they are? It turns out that this is a very difficult problem to solve, for many reasons. We will look at those reasons and the most common solution to the problem in use today, the *Public-Key Infrastructure (PKI)*.

Public-Key Infrastructures

In the last section, we looked at digital certificates as a practical method for providing trust and authentication over a computer network. We also identified a problem with simply sending certificates back and forth. There is no easy way to be sure that a certificate actually belongs to the person or company detailed in the certificate. The only guarantees that we can glean from a single certificate are (1) if the digital signature matches, then the owner of the private key created the certificate and it has not been tampered with or corrupted, (2) if the address of the sender and the common name match up, then the certificate was probably created by the owner of that address (although the address can be spoofed, leading to other problems), and (3) if the current date falls within the valid date range, the certificate is not invalid. Notice the problem here? There is no way to tell for sure that the information provided on the certificate is authentic, only that the certificate was created by the owner of the private key, the data was not altered in transit, and the certificate cannot be ruled invalid based on the current date. To this day, there is no agreement on how to best handle this problem, but there are solutions in place that provide some guarantee of trust. Without them, there would be no e-commerce, and possibly no Internet (at least as we know it).

The most common solution in place today is the *Public-Key Infrastructure*, or PKI. A PKI is not a single entity or solution, but rather the idea that a known, trusted, third-party source can provide trust to anyone who needs it. To illustrate the concept, I use the analogy of a driver's license. In common, everyday life, we are occasionally called upon to provide some proof that we are who we say we are. Whether we are withdrawing money from a bank account or purchasing alcohol, we must provide proof in order to protect ourselves and others from fraud. Our typical physical proof is our driver's license. The license has some security features built in, such as a common format, identifying information (including a photograph of the licensee), and anti-copy protection (some licenses have difficult-to-copy holograms built in). Using these features, the person requiring the proof can be fairly certain of the identity of the license holder. The reason that this trust can be extended is the fact that the license was issued by a third party, in this case, the government, that is inherently trusted (some would say that we cannot trust the government at all, but if we did not, there would be no society, and we should all become hermits!). The government, in this case, extends its inherent trust to the licensee, and the anti-copying features of the physical license back up that trust.

In the electronic world, there is effectively no way to tell the difference between an original document and a forged copy. The only way that we can be sure of the validity of an electronic document is through some type of cryptographic mechanism—generally speaking, this is a digital certificate (assuming that a cryptographically secure channel has not been previously established). Digital certificates can be generated by anyone, but a few companies provide "signing" services where a certificate holder pays for a digital signature to be generated for that certificate. The companies that provide the signing services are generally referred to as "Certificate Authorities" (CA), and they have a private key that is used to sign customer certificates. This private key is associated with what is known as a "root" certificate, which is the basis of trust in a PKI hierarchy. The root certificate is often provided in web browsers or on secure Internet sites. The basic idea is that a root certificate can be loaded from a known trusted site, providing a fairly high level of assurance that the

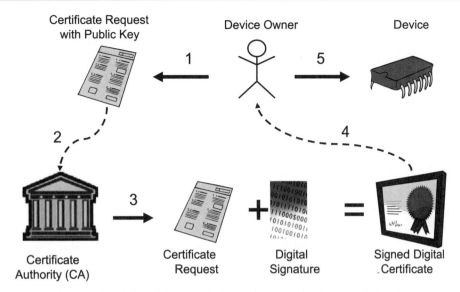

Figure 10: Digital Signature Signing Using a Certificate Authority

certificate is valid. As long as a CA keeps the private key hidden, and provides the public-key in the root certificate to the public, the PKI infrastructure works. Next we will look at how a root certificate can be used to verify an unknown but signed certificate—see Figure 10.

To verify a digital certificate, the recipient of that certificate must have access to the root certificate used to sign the certificate in question. The user can then decrypt the digital signature using the root certificate's public-key, and verify the hash of the remaining certificate data against the decrypted signature. Assuming that the CA keeps the corresponding private key hidden from the public, then the user can be fairly certain that the unknown certificate has been verified by the "trusted" third-party CA. As long as the CA is trusted by the user, then verification via this mechanism extends that trust to the unknown party.

Obviously, the trust in the CA is the most important link in a PKI chain. If you do not trust the company doing the signing, then there is no guarantee that a signed certificate has any validity whatsoever. However, CA's rely on trust, so their reputation, and hence their profits, are directly tied to the verification of certificate holders. Through traditional and physical means, a CA will usually follow up on the identity of a certificate holder before providing an authorized digital signature. The caveat here is that the CA provides only a guarantee that the certificate matches the person (or entity) providing it, not that the person is inherently trustworthy. It is completely up to the recipient of the certificate doing the verification to decide if the provider is trustworthy.

Certificate Authorities can also extend trust to other companies or entities that provide signing services under the umbrella of the root CA's trust. These companies are called "intermediate Certificate Authorities." An intermediate CA has its own root certificate that is actually signed by the root CA. Through this hierarchy, trust can be extended from the root CA to the intermediate CA, and finally to the end user. This hierarchy of extended trust is typically referred to as a "certificate chain," since the authentication forms a chain from the root CA to the end-user certificates. This chain is precisely like the governmental hierarchy from the analogy above, where the root CA is like the government, the intermediate CA is like the Department of Motor Vehicles, and the end-user certificate is like the driver's license. For better or worse, however, a CA is not an elected body, but rather a company that has established itself in the industry as a trusted signing entity. The foremost example of a root CA operating at the "governmental" level is Verisign. A quick look in any web browser at the built-in certificates will show a large number of Verisign certificates, or certificates from intermediate CA's that are covered under the Verisign PKI—see Figure 11 for a comparison with our previous driver's license example.

PKI is definitely not the only way to provide a network of trust for digital documents, but it has become the de facto standard because of the perceived trustworthiness in paying for authentication. One of the major drawbacks of PKI is that a single company or small group of companies controls the entirety of trust on the Internet, creating both a bottleneck and a single point of failure. Some security experts are of the opinion that PKI is inherently flawed for this and other reasons, and an alternative is needed. One such alternative is peer networking. A person establishes trust with someone they know, and then trusts documents sent by that person. Once a certain level of trust is achieved, then that trusted peer can

Government		Root CA
Department of Motor Vehicles		Intermediate CA
Driver's License		Signed Certificate

Figure 11: Department of Motor Vehicles vs. Certificate Authority

vouch for the validity of documents provided by people or entities they know and trust, even if the recipient does not know the sender. By keeping the number of "hops" between a sender and a recipient short, the trust can be fairly easily maintained—without a central body providing that trust. The advantages of this are fairly clear, each person controls what is trusted, and there is no single point of failure. The problem with adopting such a scheme, however, is that establishing peer networks of trust takes time, and there is a problem if the sender of a document is not connected to the recipient's network. In any case, PKI has some definite problems, and we will likely see some improvements or a replacement as the Internet continues to permeate our lives.

Wrap-Up

Hopefully this chapter has been a good introduction (or review!) of the basics of computer security and cryptography as we move now into the next chapter where we cover network communications and the built-in security features of several common protocols. Now that we have covered the basics of security and cryptography, we need to look at communications protocols and mechanisms to see what types of security are needed when sending information over a network. Sure, there is some use for security on non-networked devices, but generally speaking, security is most useful in networked applications. We can use encryption to protect sensitive information in a stand-alone device, but this is generally a much smaller issue than providing security over a communications channel. We will cover some of the security issues of a stand-alone application when we discuss physical security in Chapter 10 but the majority of the text will focus on communications security. For this reason, the next chapter focuses entirely on communications protocols and their inherent security features (or lack thereof!). We will look at several common protocols that use varying levels of security mechanisms, from the simplest hashing scheme to the most complex security suites that are required to meet the protocol spec. One notable exemption from Chapter 2 is SSL, which will be covered in detail on its own in Chapter 4.

Recommended Reading

Practical Cryptography—Bruce Schneier

Applied Cryptography—Bruce Schneier

Computer Security: Art and Science—Matt Bishop

Network Communications Protocols and Built-in Security

Embedded systems that are connected to any network, be it the full Internet or a small LAN, have to conform to the various protocols that have been designed and implemented to keep the network running—see Figure 1. Obviously, there must be communications protocols, including low-level (such as Ethernet or wireless technologies), medium level (such as TCP and IP), and high-level (HTTP, FTP). These protocols all work together to provide the vast array of services available from networked devices. Each of those levels, however, is subject to a number of different failures and attacks, and therefore, many protocols have been adapted to include a notion of security. For the low-level, security may be as simple as checking hardware MAC addresses against a known list, or as complicated as WPA and other wireless technologies. In the middle, there are protocols like the Secure Sockets Layer (SSL) that provide blanket security for higher level protocols. At the high level, there are various schemes that protect information, such as HTTP authentication. With all of these protocols, how can a simple, resource-constrained device hope to keep up?

Well, in order to answer that question, we first need to look at some of the protocols that we need to implement in order for our devices to be able to be connected to a network. Only by looking at some details will we be able to pick the protocols that are relevant to our applications—and "trim the fat" we do not need off of the protocols we do need. As it turns out, there are so many options available to us that we can pick and choose not only the protocols that we need, but also *parts* of those protocols that we are most interested in. This chapter will cover a number of protocols and technologies that are commonly used in networked systems. We will look at each from a high level first then dig into the details. The point of this chapter is to familiarize the reader with the technique we will use later to choose security options for a particular application. Essentially, we will look at the intended use of a protocol or technology, then what makes that technology work, and finally, if and how we can adapt that technology to suit our requirements. For now, we will only look at the protocols from a standard PC-level view. Chapter 5 will cover the specific challenges related to using these technologies in a constricted resource environment.

The following is meant to be either an introduction to the material or a refresher for more experienced readers. It is not intended as a complete reference for the protocols discussed,

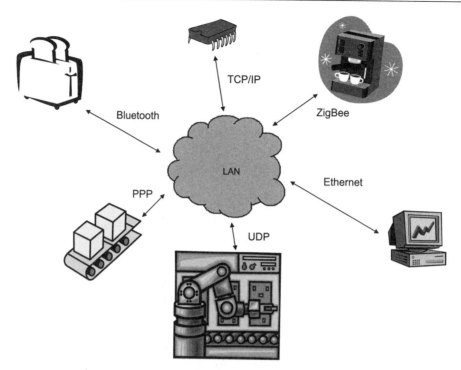

Figure 1: Embedded Devices on a Local Area Network

but rather as a starting point for research. We will break down each technology to get a good idea of how it will be useful for us, leaving the mundane details to other references. This chapter will provide a basis for understanding the role of security in network communications protocols using protocols primarily designed for wired communications. We will use this as a springboard into the material on security protocols in the next chapter. We end the chapter with a short discussion on wireless protocols, but due to the complexity of the security for wireless communications (and the fact that security is a major feature of all wireless protocols), we will leave the detailed discussion of wireless security for Chapter 6, after we have had some time to cover security protocols (Chapter 3), the Secure Sockets Layer (SSL, Chapter 4), and the detailed discussion of security issues for resource-constrained systems (Chapter 5). The reason we push off the wireless discussion is that the intervening chapters all cover material that is incorporated into the common wireless standards (for example, 802.11 Wi-Fi relies on SSL for many of its built-in security features).

Low-Level Communications

We will start our quick tour of communications protocols with the lowest network level. These technologies are often hardware-based, very low-level software, or a combination of

Application	
Transport	
Network	IP, ARP, IPSEC
Link	Ethernet (software), 802.11 MAC, PPP
Physical	Ethernet (physical), 802.11 radios, RS-232

Figure 2: The Network Stack Model with Example Protocols

both (usually assembly code or some form of driver are needed to communicate between the hardware and higher-level protocols). This category includes wired protocols such as PPP and Ethernet, as well as wireless technologies such as Wi-Fi and Bluetooth. Each has specific security requirements and each deals with security in its own way (which may be to have no security options at all). We will look at four of these protocols and discuss what features they have to support robustness and security. See Figure 2 for an example of the network stack with some of the protocols we will be covering in this chapter.

Point to Point Protocol (PPP)

PPP is a relatively old communications protocol, described in 1994 in RFC 1661,[1] and was designed to provide connectivity over serial hardware channels. PPP was originally developed to allow higher-level protocols to utilize these serial channels in a consistent manner. This protocol, though losing out to newer, faster technologies, is still used widely for embedded systems due to the fact that simple serial hardware is much less expensive than the hardware that some of the newer standards require. PPP consists of a few protocols designed to establish the serial link, encapsulate the higher-level data, and to control each of the high level protocols that can be used. Many different high-level protocols are compatible with PPP, but since we are discussing Internet security, it is reasonable to restrict our discussion to cover only the Internet Protocol, or IP. We will see in a minute how IP and PPP work together, but first we will look at the link establishment and discuss what it means for our embedded applications.

PPP, as with many low-level protocols, is designed to be the connection between the network hardware and the application. Link establishment in PPP is controlled by the Link Control Protocol, or LCP. The LCP divides the link establishment procedure into 4 distinct steps:

---

[1] Request For Comments (RFC) is the format used to describe most Internet protocols. RFC documents are controlled by the Internet Engineering Task Force, or IETF.

1. Establish the serial link using the hardware.

2. Optionally, test the quality of the link to determine if the hardware can handle the communication level desired.

3. Negotiate and configure the higher-level protocol for transmission.

4. Terminate the link and release the hardware.

To illustrate how we can adapt the protocol for our needs, let's look at the steps of the LCP. Obviously, we have to have a hardware connection in order to communicate with the remote device. However, the link quality test is an optional step; leaving it out may result in subpar communications, but it will definitely save on code size, and the link can be established faster. If we were trying to use PPP on an 8-bit CPU with only 64KB of program and data space (combined), this is a place where we can definitely cut corners to save on precious space. The tradeoff here is that while we obviously gain more space for our program and a small performance advantage, we lose some guarantee of robustness, and therefore, security. Imagine that an attacker could interfere with your communications by generating a high amount of electronic noise near the wire used for communications. Without the link quality test, you will not be able to determine if the link can handle the connection you need. The attacker then can hinder your application without having to cut a wire or hack into the system.

This is the type of tradeoff that we will focus on in developing secure applications that need to fit onto our resource-constrained devices. On one hand, we have the strict size requirement, on the other a potential security risk. Depending on the requirements of the application, it may be more desirable to remove the optional step and save precious space. We will look at some of these requirements so you can understand how to look for these types of options when dissecting security and communications protocols. For now, let's continue looking at PPP and see what we can do with the communication protocol itself.

PPP is an inherently configurable protocol, allowing for many implementation options. For each network-layer protocol, there is a corresponding Network Control Protocol (NCP) that allows the protocol to utilize PPP as the link-layer transport mechanism. Each NCP is designed to provide the correct functionality to allow PPP to transport the higher-level packets. This functionality includes any security protocols inherent to the higher level protocols, such as IPSEC (security for IP). The selection of NCP's to support is another option we have to conserve code space. If we know that our application will only need to support one higher-level protocol (such as IP), then we only have to implement the functionality specific for the NCP for that particular protocol. We can tailor the implementation to fit the protocol, and since we do not have to support the other NCP's, we can simply reject any protocols that we do not support. We need to be sure that the rejection

Byte positions

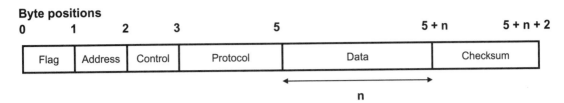

Figure 3: PPP Structure

mechanism is robust, however, since if the mechanism is not well behaved, it could set up a situation where a denial-of-service or buffer overflow might be possible. For example, if the LCP on the remote end (asking for a link to the device) sends a protocol request mechanism that is very large, it can overflow local buffers, allowing a user on the remote end to crash the device or run arbitrary code (depending on the local implementation). If the rejection mechanism is slow, or if the implementation allows multiple connections from a single remote device with no limit between retransmissions, a remote device could send a flurry of connection requests, effectively preventing the device from serving legitimate requests. This is an example of where robust design and programming can make the device more secure.

PPP has its own security mechanisms that we can use to authenticate connection requests, allowing the implementation to protect the device from unauthorized use. The security mechanisms supported by PPP are password authentication and a challenge-handshake. Again, we can choose to support either of these mechanisms. The password mechanism will have a simpler implementation, since the challenge-handshake will require additional states in the PPP state machine to handle the additional messages. However, we may also choose not to support any PPP security, instead relying on the higher-level protocols to provide security for the application. Depending on the application, the no-security option may be more desirable. If the network is not secure, sending a password would allow anyone eavesdropping on the network to read the password.

The challenge-handshake protocol, though more complex, is also more secure than the password protocol. The common challenge-handshake protocol for PPP is defined in RFC 1994 (written in 1996), and is referred to as the Challenge Handshake Authentication Protocol, or CHAP. CHAP provides decent security for devices with a previously defined trusted relationship, but since it requires shared secret keys (cryptographic keys stored on each end), it is not practical for general-purpose security (connecting to arbitrary remote systems). Without an established relationship, the secret for the challenge mechanism must be sent plaintext over the network—which is obviously not secure at all. Whether or not such a relationship can be established should be a factor in deciding whether or not to support the authentication protocol.

As was mentioned previously, PPP is an older technology, but is still popular on smaller embedded devices because of its ability to use inexpensive networking hardware. However, current technology is moving toward newer, more complex low-level protocols, and these new technologies have more security options, but also more security challenges. Your application requirements will determine whether PPP or one of these other technologies should be used. Next we will look at another applicable technology that has gained wide-spread popularity and is practically the only technology used for Local Area Networks (LANs). This technology is practically a household word—Ethernet.

Ethernet and ARP

Where PPP is predominantly a software protocol, Ethernet is a combination of software and hardware. Although Ethernet is even older than PPP, it has gained widespread acceptance and is the de facto standard networking technology for LANs. Developed originally in the late 1970s and early 1980s by Digital Equipment Corporation, Intel, and Xerox, the Ethernet standard describes both a physical hardware layer and a higher-level data encapsulation layer. Ethernet has been an evolving standard, and has kept pace with new developments in networking technology. The standard is now part of the IEEE 802 networking standards collection and newer versions support some of the highest speed hardware, capable of transmitting gigabytes of data each second. This Gigabit Ethernet is still relatively expensive since it requires expensive and powerful hardware. For this reason, we will limit our discussion of the protocol to older versions that are within the capabilities of resource constrained systems having tight economic limits. The current widespread version of Ethernet that will likely be familiar to the reader is 10-Base-T, capable of up to 10 mega*bits* per second (Mbps). Likewise, 100-Base-T is capable of 100Mbps.

Ethernet is characterized by a physical layer that utilizes internationally assigned unique Medium Access Control addresses (MAC addresses) that are simply large numbers that are globally unique to a particular device.[2] The data-link layer described by the Ethernet standard defines a *frame* that encapsulates both the MAC address and the data being transported. Shown in Figure 4, the Ethernet frame is used by the hardware to control where the data is going on the physical network. There are several technologies that may be used to implement the physical layer, but this discussion is beyond the material covered in this book.

However, there is part of the Ethernet stack that is of interest to us, and that is how the addressing of each device is resolved—each device can have several addresses. These addresses include the hardware address (MAC), IP addresses, and TCP ports, among others.

---

[2] MAC addresses are controlled by a single entity, the IEEE, so they are responsible for making sure that assigned addresses for commercial products are all unique.

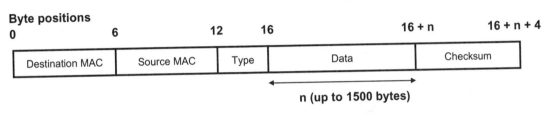

Byte positions

n (up to 1500 bytes)

Figure 4: Ethernet Frame

The software layers need to be able to resolve the physical address of the destination device requested by an application so that the data can be communicated to the right receiver. The primary way this is done is to use the Address Resolution Protocol, or ARP. ARP is responsible for associating hardware addresses with higher-level protocol address, primarily Internet Protocol (IP) addresses. Described in RFC 826, ARP facilitates the translation of higher-level protocol addresses, such as Internet Protocol (IP) addresses, into Ethernet hardware MAC addresses, and the reverse, associating a hardware address with a particular higher-layer address. ARP works by storing a table locally that contains the information about all of the known devices on the Ethernet network. This ARP table is updated only when a device is added to the network or a device attempts to contact a device not already in its table.

ARP works by broadcasting a message to all the devices on the Ethernet network, asking if a particular device has the higher-layer protocol address and can speak the right protocol. Once the target device receives the message, which contains the sender's hardware address, it updates its own ARP table and sends a reply back; this causes the sender's table to be updated. In this fashion, the ARP table does not have to be constantly updated, sending flurries of packets over the network whenever a connection is desired. The sender and receiver can contact directly using the stored hardware addresses.

One of the first things to notice here is the inherent trust of the underlying hardware to provide the correct information. It is fairly trivial to dupe a device by copying the MAC address and providing it when the ARP requests are broadcast. In fact, ARP is utterly insecure on its own—the protocol specification does not include any type of authentication mechanism. Attacking ARP is quite easy, and it is typically done using what is known as *ARP Cache Poisoning*. This term refers to the introduction of improper ARP table entries by a malicious user to gain certain privileges on the network. For example, a hacker could insert his machine into a network, and by sending out carefully crafted ARP requests and replies, can cut off the network by changing an ARP table entry for a router to an entry pointing to a nonexistent machine. The devices on the network happily send packets to the address in the table, blissfully unaware that those packets are going nowhere.

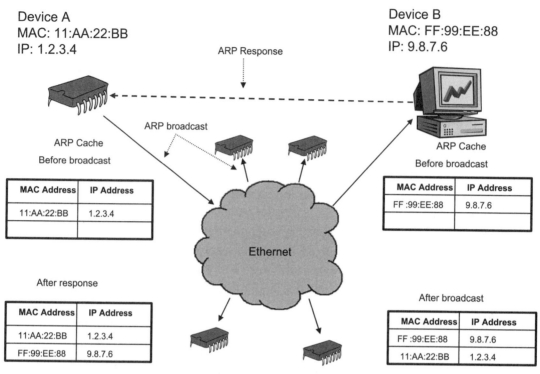

Device A
MAC: 11:AA:22:BB
IP: 1.2.3.4

Device B
MAC: FF:99:EE:88
IP: 9.8.7.6

ARP Response

ARP broadcast

ARP Cache
Before broadcast

MAC Address	IP Address
11:AA:22:BB	1.2.3.4

ARP Cache
Before broadcast

MAC Address	IP Address
FF :99:EE:88	9.8.7.6

After response

MAC Address	IP Address
11:AA:22:BB	1.2.3.4
FF:99:EE:88	9.8.7.6

Ethernet

After broadcast

MAC Address	IP Address
FF :99:EE:88	9.8.7.6
11:AA:22:BB	1.2.3.4

Figure 5: ARP Broadcast Steps

Other possible exploits include overwhelming a network switch, allowing the hacker to monitor all network traffic by bypassing the switch's security features (some switches, which are designed to be "smart" by being aware of connected machines drop into a "dumb" mode when too much ARP traffic is received—the switch effectively becomes an insecure hub, allowing full access to all network traffic). Probably the most insidious attack is a man-in-the-middle attack, where the hacker inserts a machine into the network between the victim device and a router. The device thinks the malicious machine is the valid router (since it relays packets to and from the router), and neither end of a connection will be aware that the machine is listening.

Fortunately, ARP requires local network access, which is easier to restrict than in the general Internet. Placing your device behind a firewall and router will lessen the threats posed by ARP attacks. Also, using higher-level security, such as SSL, will help to assure that even if the network ARP tables are breached, the attacker cannot get at any useful information. The problem for embedded devices, however, is that the attacker may have access to the network, since the device is likely in the field, not behind locked doors in a closed environment. How can we hope to protect our device from ARP attacks when we may not have complete control over the physical device at all times? It turns out that there are a few things we can do to improve the situation for ARP.

The primary method for protecting an ARP cache is to make it *static*. In other words, the ARP table is made unchangeable, and the device is restricted to communicating only with those devices already in the table. Not allowing the table to be updated prevents ARP cache poisoning, since there is no mechanism present to update the table. This presents an additional bonus for a resource-constrained system, since using a static table allows us to remove the ARP handling code, freeing up more space for the application.

Unfortunately, ARP is essential to most network communication, so it will likely be required for most applications. Other than the static ARP cache trick, there is very little we can do to directly protect the device. However, this does not mean that all is lost, since we can use security mechanisms at higher network layers that will protect from the man-in-the-middle and eavesdropping attacks, as mentioned above.

The denial of service attack, where the device is cut off from the network, can be prevented using a static ARP table—but what do we do if we need a dynamic ARP table, assuming that the device is on a constantly changing network and needs to communicate with many devices? In this case, we have to deal with ARP cache poisoning in a different manner. One possibility would be to use a periodic request at a higher layer to assure that the device is still connected to the network. As an example, let's look at a hypothetical system, where our device is plugged into a corporate network and used to log when the door to a high-security area is used. Now, let's assume that the network is dynamic, with all IP addresses being assigned using a protocol such as DHCP. Once a month, the network is shut down for maintenance, and the subsequent reboot reassigns all the addresses. Our device needs to relay information to a central location where the logging is stored, but the IP address may change due to the maintenance shutdown (assume that the central machine's address is available through DNS or some other higher-level service). This presents an obvious problem, since the device now needs a dynamic ARP cache, thus opening the device to denial-of-service attacks. In general, denial-of-service attacks are problematic, since it is very difficult to distinguish between heavy traffic or unusual conditions and an active attack. However, in the case of our example application, we can use a simple trick to assure that the device remains connected to the network.

The trick for preventing DOS attacks on our hypothetical device involves the central machine and the device each communicating a periodic "ping," where each side issues a request and expects a reply. If we set our duration between pings to a certain time, we can be sure that the device will not be cut off from the network for longer than that period of time. For the sake of our example, let's set the ping period to one minute. This is long enough to assure that we do not have a serious impact on network traffic but short enough

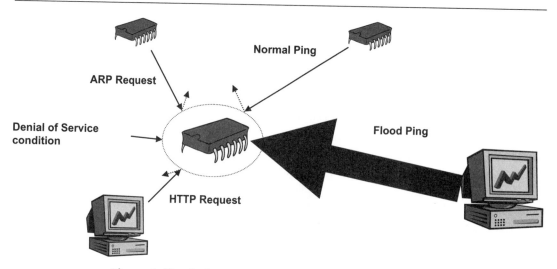

Figure 6: Flood Ping Can Bring a Device to a Grinding Halt

that an attacker would not have much time to do anything too serious before the breach was detected. Unfortunately, this sets up another type of DOS attack, where the attacker could send a flood of ping requests to the device, rendering it effectively useless. At this point, it would be up to the network administrator to identify the unusual traffic pattern and hopefully thwart the attack.

All in all, ARP is very important as it forms the basis for dynamic Ethernet networks, but it is, unfortunately, dangerous due to its inherent security pitfalls. As long as we know this, however, there are a few tricks we can employ to mitigate the worst possibilities, and hopefully handle the rest at a higher level.

Transport and Internet Layer Protocols

Now that we have discussed the lower-level protocols (not including wireless, which we briefly cover at the end of the chapter, and more in-depth in Chapter 5), we are now going to take a look at higher-level network protocols, specifically focusing on the transport layer. The transport layer is the most common lower layer that an application designer will deal with. The lowest level protocols that we have already covered are usually buried deep within hardware-specific driver code, or in hardware itself. It is in the transport layer that the application engineer first has significant control over the network settings and how his or her application will interact with the network stack. This is also the area where we can truly begin to limit features to keep our code size in check, since transport-layer protocols are more often written in a high-level language (lower levels are often coded in assembler for speed) and the libraries are more likely to be available to the programmer. For this

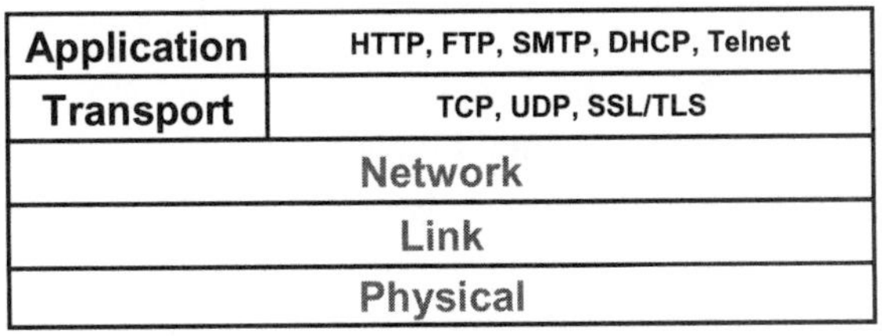

Application	HTTP, FTP, SMTP, DHCP, Telnet
Transport	TCP, UDP, SSL/TLS
Network	
Link	
Physical	

Figure 7: Transport and Application Layers in Network Stack

reason, this is also a place to focus on network security in an application, and as we will see in later chapters, many existing security protocols, such as SSL, reside here. Figure 7 shows where the transport and application layers reside in the network stack.

We will start our discussion with two protocols that are almost always lumped together due to their widespread applicability for Internet applications. These protocols, called the Transport Control Protocol (TCP) and the Internet Protocol (IP), form the software basis of the Internet, and their use is so widespread that it would be very rare to find a network-capable application that did not support them. Together, these protocols are typically referred to as TCP/IP due to the fact they are rarely used separately. The other protocol worth mentioning in this section is the User Datagram Protocol (UDP), which can be thought of as "TCP-lite," since it works on the same basic principles as TCP but does not have the reliability guarantees provided by TCP. We will see that UDP is useful for many applications that do not strictly need the reliability features of TCP.

Transport Control Protocol and the Internet Protocol (TCP/IP)

TCP was one of the earliest "reliable" protocols, in that the application does not have to worry about the fragmentation of information. To the application, TCP is like a pipe in that what goes in one end will always come out the other end in the same order it entered. TCP utilizes several methods to keep data in order and to assure all data is sent and received, and as we will see, this property is extremely important to the security of a networked application. TCP in and of itself does not provide any encryption or explicit security features, but the reliability and robustness of the protocol make it the perfect foundation for a transport layer security protocol. Indeed, security protocols like SSL, TLS, and SSH all rely on the inherent reliability built into TCP in order to function properly. We will cover this in later chapters when we discuss those protocols, but the basic idea is that the encryption used in those protocols requires that all the data being passed over the network (in its encrypted form) must arrive, and it must arrive in the order it was sent, or the decryption operation will fail. Right now we will look at the features that make TCP reliable and how they make the protocol a good basis for secure network communications.

To understand why TCP is reliable we need to look at what would make a network unreliable in the first place. TCP generates packets of information that contain fragments of the data being sent. These packets may be routed through a dozen or more different devices (ranging from simple servers to switches and routers) over different physical media (copper wire, fiber-optic cable, or through the air as radio waves—see Figure 8). Each machine has to decipher the packet header information (including the target machine address) and send it to the next device in the path to the target. At any of these points, a device may fail to propagate the packet for any number of reasons, such as a power failure or a broken network connection. When traveling over physical media, another set of problems arises, such as external noise or a cable being severed (as sometimes happens to fiber-optic cables near construction sites). All of these problems are simply artifacts of sending information over a network. Add to that a malicious attacker and you can have some serious problems, since any or all of those problems can be exploited to perform an attack.

So how does TCP assure that all your data gets to where it needs to go? Well, first of all, the protocol assumes that the underlying network layers are unreliable. In order to combat

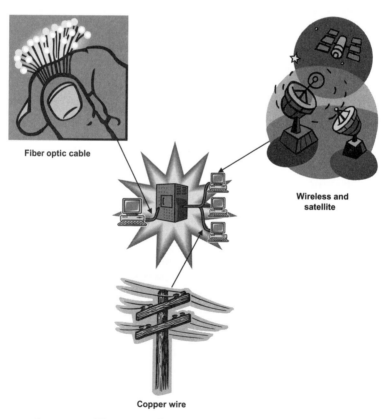

Fiber optic cable

Wireless and satellite

Copper wire

Figure 8: Different Media Types Used for Communication

this assumption, several mechanisms are used. First, when a TCP connection is established, the protocol does not simply start sending information to a target address. It performs what is known as a handshake, where several packets of protocol-specific information are exchanged between the sender and receiver before any application data is sent. In this way, the sender can be assured that the target does indeed exist and that there is an available connection to that target. After the connection is established, TCP uses two primary methods of keeping the information intact. First, it utilizes a counter that is incremented for each packet sent and included in the packet header. The receiver then can automatically tell if there is missing information (a missing packet will mean a number is missing in sequence, which the receiver can easily determine), and it can easily reassemble the packet data into the order in which it was sent by putting the data in the order of the packet counter. If a packet is missing out of the sequence, the receiver will send a special packet back to the sender indicating which packet is missing and telling the sender to retransmit the missing packet.

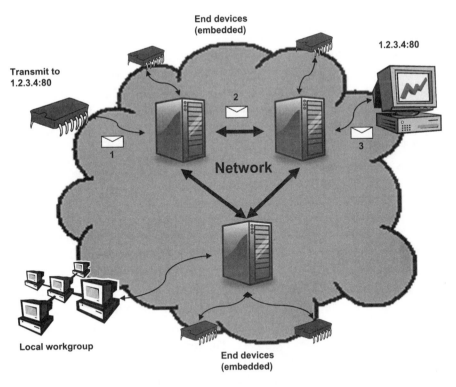

1)   Packet routed to server
2)   Server routes packet to recipient's server
3)   Recipient server forwards packet to recipient

Figure 9: TCP/IP Packets in the Network

The other method TCP uses to provide reliability is the idea of a dynamically-sizable "window" that the protocol can adjust to compensate for network traffic or other problems. For example, if there is a lot of traffic on the network, and larger packets are being lost at a high rate, TCP may shrink the window to allow less information to be transmitted, increasing the probability that the information will make it to the receiver unscathed.

TCP is almost never seen alone, as it is most frequently used with the Internet Protocol. IP provides the backbone functionality of addressing on the Internet, and even basic computer users have likely heard of IP addresses (though they likely do not understand what they really are). IP follows a similar model to the telephone system, in that every device on the network (which does not necessarily need to be the Internet—IP can be used on private networks as well) is assigned a unique number. IP even has a concept of "area codes" that is analogous to the geographic designations used in telephone networks, but the "area code" in an IP address is in reality an arbitrary assignment of a particular range of addresses to a particular organization. Within that range, the organization "owns" all the addresses and can use them for various purposes. In local networks the concept is a little more useful, in that an organization can define "subnets" which allow the network to be partitioned into manageable units.

IP is a fairly simple protocol, and it mostly relies upon the underlying network hardware and the properties of TCP to provide reliability. However, a security protocol has been developed specifically to thwart attacks on the IP protocol. Called IPSec, it essentially allows "virtual" networks to be set up across a public network. Called Virtual Private Networks (VPN—see Figure 10), these provide a higher level of security that is invisible even to the TCP layer. We will look at IPSec and the concept of VPN's in a later chapter when we discuss security protocols.

User Datagram Protocol (UDP)

TCP is a relatively heavyweight protocol due to the extensive logic required to provide reliability. For some applications, the extra overhead of TCP is undesirable or simply unacceptable. Examples include streaming audio and video applications, where the fastest throughput available is needed to provide acceptable service. The User Datagram Protocol fills this niche, since it is a simple protocol without any of the reliability guarantees of TCP. UDP packets, called *datagrams,* are simply sent to a target address with essentially no guarantee that any will arrive—see Figure 11. The reason UDP works at all is that the underlying network (usually consisting of IP over some low-level protocol) is for the most part reliable. Applications using UDP must be tolerant of dropped or out-of-order datagrams, as the protocol does not catch these errors. The inherent lack of reliability in UDP makes it unsuitable for most secure applications, but it does have a useful property that TCP does not—multicasting. Multicasting allows a message (in the form of a datagram) to be

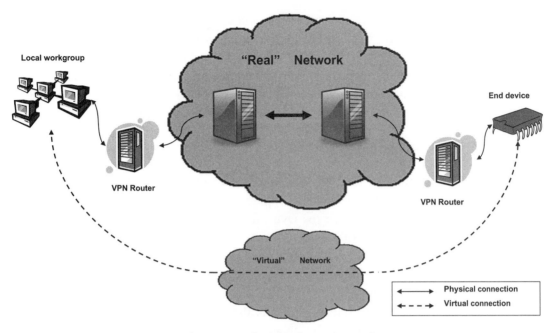

Figure 10: Virtual Private Network

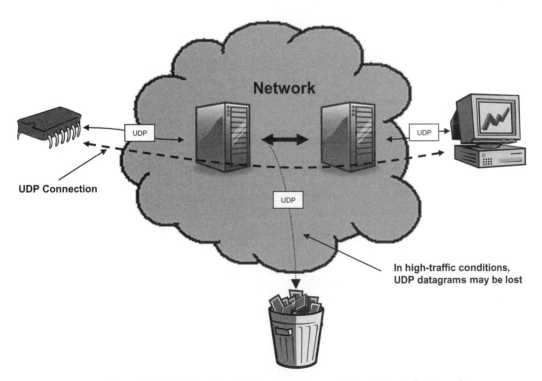

Figure 11: UDP Is Unreliable—Datagrams May Be Lost in Transit

broadcast to a range of addresses. Since UDP does not guarantee any of those addresses is reachable, this is a perfectly acceptable feature. To implement multicasting in TCP would be difficult, since every address in the range must be checked by the protocol, which would create a strain on the network resources. Multicasting is not inherently secure whatsoever, but it would be possible to write a secure application using UDP multicasting, as long as the application handled all the necessary security—encryption, ordering of data, and so on.

Other Network Protocols

We have looked at the most familiar low-level protocols so far, but there are a whole host of other protocols that an embedded systems engineer may be interested in or required to implement. Among these protocols are the Dynamic Host Configuration Protocol (DHCP), the Simple Network Management Protocol (SNMP), and subprotocols like the Internet Control Message Protocol (ICMP) and the Internet Group Management Protocol (IGMP), which are components of the IP protocol suite as seen in Figure 12. There are a variety of other protocols that you may be familiar with, such as the Simple Mail Transfer Protocol (SMTP, or email), File Transfer Protocol (FTP), or the Hypertext Transfer Protocol (HTTP). We will not discuss these protocols in this chapter in order to focus on the low-level

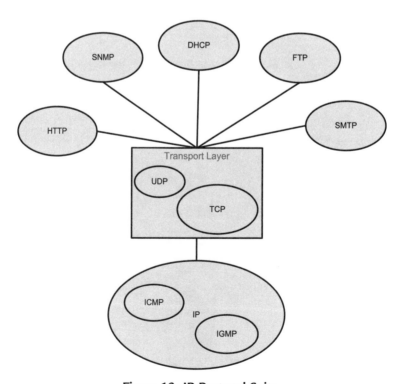

Figure 12: IP Protocol Suite

protocols that form the basis of networking, rather than protocols that utilize networking. We will take a look at the high-level protocols in Chapter 7, when we discuss the client-server model of networking.

Moving on to our other protocols, let's first look at DHCP, since it is one of the most popular ways to dynamically configure a device's networking stack. DHCP allows devices to connect to a network without having been previously configured for that particular network. Once the low-level protocol establishes a connection (physical level, such as Ethernet), a DHCP server provides all the IP configuration information for that network to the connecting device. As long as the device has DHCP enabled, it can use that information to start communicating on that network. DHCP obviously has some ramifications for security, since there is a lot of power in providing configuration information. One of the most obvious potential issues is the assignment of the same IP address to two distinct devices. This conflict is not handled very well by IP, since it will simply connect to the first device that responds to that particular address. There would also be a problem with ARP, since there would be two separate hardware addresses for a single IP address. Fortunately, if implemented correctly, DHCP will guarantee that no two devices will share a specific network address (see RFC 2131, DHCP). Unfortunately, this is only one of the ways that DHCP could be exploited, and there are a number of other vulnerabilities that make DHCP a target for attackers.

DHCP does not provide any authentication or authorization when configuring a new device, so an attacker could connect to the network and launch an attack on any of the devices on that network. An attacker could set up a DHCP client, which is a device that requests new addresses, and setup the device to drain the pool of available addresses from the DHCP server (the device providing the network configuration information). The client could also use up all the available bandwidth, thereby causing a denial-of-service problem. Worse yet, if the network is not configured to detect improper DHCP servers, an attacker could insert a device that acts as the DHCP server that could then corrupt the integrity of the network by providing falsified configuration information.

The DHCP protocol does not specifically provide any security mechanisms and it is built upon IP and UDP, both of which are considered insecure protocols, so anyone using DHCP must be careful to use other mechanisms to provide security. Though extremely useful, DHCP can be the root of a number of problems, so when designing an application, you must be sure to determine whether DHCP is strictly necessary and disable the protocol if it is not. You should also check to see that the DHCP logic is not included in your application to save code space.

If DHCP is required, the application should utilize a higher-level mechanism to authenticate new clients, and the network should be designed to authenticate new DHCP servers in some

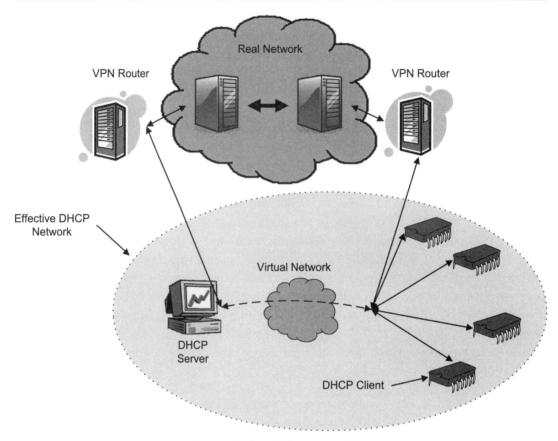

Figure 13: DHCP Used in a VPN

way. If more security is desired, RFC 3456 describes using DHCP to configure IPsec connections. This has the benefit of the security being transparent to the application, as it would be under normal IPsec operation as seen in Figure 13. As you can see, DHCP is not really all that exciting. Its use is fairly common, however, and it is important to know its limitations.

Why All the Protocol Analysis?

Before we move on to the wireless technologies, we will discuss the rationale behind our brief look at the different protocols and how they do or do not provide security. The point of covering these protocols here is to provide a basis for the upcoming chapters—it should now be clear that security is not an inherent part of the network or the protocols of which it is composed. Instead, security is an add-on feature, and as such, cannot simply be ignored. It would be nice if all the networking protocols included some reliable security, or (even

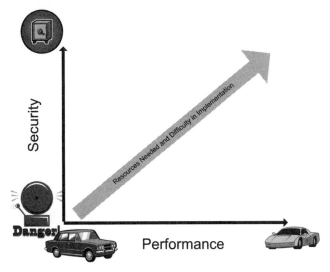

Figure 14: Security vs. Performance and Ease of Implementation

better) were inherently secure in their designs. The problem is that many of the protocols were designed for performance, and security is almost always at odds with high performance (see Figure 14). As a result, security is forgotten, and it is up to the end engineer to tack on the security at the last minute.

We have spent the past few sections doing brief dissections of a couple of the most common protocols to introduce the mindset we will take through the rest of the book, as well as to emphasize the point that security is not something that just "happens" to an application but is something that must be deliberately designed and carefully integrated into all parts of an application, starting at the lowest protocol levels. We do not cover all the protocols because that is not the point of this book (it would be enough to fill several books anyway). Instead, the goal is to pick apart an application, get up to your elbows in protocols, and question *everything*. If you need security, you need to be willing to look at parts of the system that others would be quite willing to ignore (certainly the hackers won't ignore those parts). That being said, it can be difficult to do all of that work alone—the more eyes the better, and don't keep your code a secret (it is almost guaranteed to not be secure and hiding it won't keep the bad guys out).

Moving forward in the text, keep an eye open. Nothing here is gospel truth—it is just a guide to point you in the right direction. Hopefully the crash-course in protocol security will help you to keep an open mind about security permeating the entire application. With that in mind, we can turn to more interesting things—wireless technologies.

Cutting the Wires

We have so far looked at all the "wired" protocols, since they have a conspicuous lack of security options. For the most part, the wired protocols, unless running over a public network, do not really need all that much security, except in the most critical applications. These protocols all rely on an inherent property of the network itself for basic security—wires are easy to secure. This property does not become apparent until you try to remove the wires and broadcast information using a radio. Suddenly, every hacker, black CIA helicopter, and alien invader in the vicinity can read everything you send without you knowing about it. If someone were to try to eavesdrop on a wire, they would need physical access to that wire. Without the wire, all they need is an antenna (and maybe a dish to amplify the signal). See Figure 15 for some example threats to wireless communications.

One of the primary reasons wireless technologies have not been as prevalent so far (even though the radio predates the Internet) is that wireless communication is hard—everything from cell phones to sunspots to microwave ovens to Grandma's pacemaker emits some kind of radio noise—see Figure 16. If any one of those things is producing loud enough noise on the frequency you want to broadcast on, it can be extremely difficult to pick out your specific message (this also brings up an interesting point about denial-of-service attacks—how can you know if you are experiencing a DOS attack or a bad day of sunspots?). Slightly less of a factor, but nonetheless a contributor to the problem, is the fact that security is also hard. As we said, all it takes is an antenna for someone to listen to your broadcast, but tapping a wire is at least a little more difficult. For this reason, security is basically an absolute necessity for any type of wireless communication technology. Until the security

Figure 15: Wireless Communication Threats

Figure 16: Radio Noise Sources

capabilities of the systems employing wireless communications caught up with the wireless technology, it could not progress too quickly. Of course, when wireless technologies started to catch on, it started in the high-end PC market. Only now, several years later, is that technology percolating down into the inexpensive embedded control industry. Starting in the next section and continuing into the following chapter, we will look at some of the most common wireless technologies and the security implications of using those technologies. Armed with the analyses of the wired protocols from the previous sections, it should be apparent that securing wireless applications can be both easier and more difficult than securing their wired cousins.

Wireless Technologies

For many years, wired network hardware has dominated the Internet landscape. Recently, however, technology for wireless communications has reached the point where it is reliable and powerful enough to be universally adopted. With the dawning of a wireless revolution, there will be a similar revolution for embedded technology, as wireless capability will allow for applications that never were possible before. The current standard leading the way is actually a collection of wireless networking protocols, similar and related to Ethernet technology. By far one of the most mature wireless standards, Wi-Fi (802.11a/b/g and other variants) is the dominant standard for computer systems. Cellular technology is bigger, but less Internet-friendly, and generally used for telephone communications. However, cellular is quickly making inroads to the embedded space, so we will cover cellular communications in another section.

Wi-Fi should be familiar to most readers as the wireless communications protocol used to network laptops, PDA's, and other portable devices. Wireless technology is less mature than its wired counterparts, primarily due to the technical complications associated with wireless communications over hardwired solutions. However, given the many years of experience with wired networking, Wi-Fi was developed from the ground up to include security. It is obviously easier to drop in on a wireless connection than a wired network, since no physical connection is required for the former.

Wireless technology predates the modern computer by many years. Radio was the dominant form of mass communication before the advent of the television, and it still plays an important role today. The road for wireless computer communications has been relatively slow as compared to the rest of the industry, since there were new concerns that had to be addressed. A modern computer or embedded device requires a *reliable*, robust communications medium, since even a single bit off can cause major problems for an application. With radio, the information is analog, some information can be lost from an audio stream and usually no one notices. This switch from analog to digital communications has meant that the software behind the wireless technology has to compensate for many more factors than for wired technologies. Signal degradation, interference from potentially infinite sources, and the lack of control over the transmission medium (the air) are all factors that must be handled somehow. Some of these issues can be handled by the appropriate hardware design, but a good part of the "magic" of Wi-Fi happens in software.

As has been mentioned before, a secure system is robust. The advancements for Wi-Fi that provide the necessary requirements for digital communications also aid in securing the system. Wi-Fi has support for several built-in security technologies as well, and these should be utilized in any application supporting the protocol.

The Wi-Fi protocol, as defined in the IEEE 802.11 standard, is designed to mimic the properties of a standard "wired" protocol. For example, in a wired network, there can be some level of physical security applied to the hardware transmission layer. With Wi-Fi, this is obviously not possible, since anyone in the general vicinity of the transmitter can receive information from the network, because the wireless access point acts as a broadcast radio—akin to the FM radio you listen to in your car. Wi-Fi is trivial to eavesdrop on, so how does the standard compensate for this blaring security hole? The standard defines an authentication mechanism which is appropriately named the Wired-Equivalent Privacy, or WEP. WEP is a shared (symmetric) key system in which the keys must be manually (i.e. typed in) transferred to a device that will connect to the network. Note that WEP does not protect the entire transmission; its job is simply to provide the same level of security as would be found in its wired counterparts (well, probably not really, if you want to get technical—see Chapter 6 for more information on the problems with WEP).

WEP keys may also provide authentication services in addition to the privacy service described above. This satisfies a second requirement to make the wireless network behave as if it were a wired network. This prevents unauthorized access to the network (e.g., a malicious hacker logs on to the network and generates enough traffic to effectively make the network useless—another DOS attack). The authentication itself is simple—since the keys are shared in advance, simply acknowledging the actual key (via an encrypted channel) is sufficient to authenticate the requesting device. Note again that WEP authentication, like the privacy provided by WEP only serves to bring the inherent security of the network up to the general level of a wired equivalent.

So Wi-Fi is wireless, about as secure as a wired connection, and is built from the ground up to support various authentication and privacy mechanisms. How do we fit all of this onto a tiny system with limited resources? Thankfully, the designers of the Wi-Fi standard left some wiggle room in the specification that we can use to our advantage. First, there is actually no requirement for authentication and privacy. Sure, this may violate assumptions used by the higher-level applications and layers, but what if those assumptions could be eliminated? For example, let's assume that our device is part of a stress monitoring system in a crane at a copper mine. The tangle of cables needed to wire up a network to connect the device to the crane arm are just not feasible, so the application designer went with a wireless connection, as seen in Figure 17.

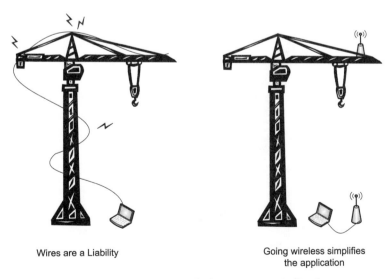

Wires are a Liability   Going wireless simplifies
                                        the application

Figure 17: Wired vs. Wireless Crane Application

The strain gauge continually monitors the stress on critical parts of the crane and alerts the operator if anything is out of the ordinary. In this application, we do not care if someone can read the information, since all it will tell them is how stressed the crane arm is (it may be possible that *someone* could use that information for nefarious purposes, but it is difficult to see how). However, we do not want just anyone to log on to the network, since they might be able to overwhelm the system and cause the crane to shut down or behave erratically. In this example, we only need some type of authentication, not any type of privacy. By removing the privacy requirement, we do not need to implement a full cryptographic system on the device, freeing up more space for the actual application. Of course, the cryptography may be implemented in hardware, but this not may be cost-effective if thousands of devices were to be deployed.

So we were able to eliminate the privacy requirement, but now we have the problem of authentication. For this, we can use a simple challenge-response mechanism. For example, both the device in the field and the network wireless access point have shared cryptographic keys, they are just not used for encrypting anything. Instead, the access point can provide a cryptographic hash of a secret key and some information, usually a time stamp and some unique value. The field device uses its secret key, which is a copy of the server's secret key, to produce a new hash that includes a password and the information from the server. The server can then generate a hash locally with the expected password and compare it to the one received from the client. The timestamp and unique identifier help to combat replay attacks (the malicious device sends a copy of the client's response to attempt to gain access to the server), and put a time limit on how long it will accept any response without restarting the process. Note that this is essentially the same challenge-response algorithm

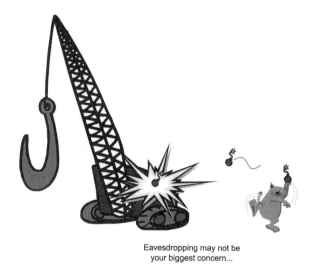

Eavesdropping may not be
your biggest concern...

Figure 18: A Really Bad Day at Work . . .

developed for HTTP and described in RFC 2617 (we will cover HTTP authentication later). As such, it suffers from the same problems as the HTTP mechanism, and should not be relied upon as a complete security solution. However, for the application illustrated in our example, the challenge response should be sufficient, since it will prevent most intrusions without increasing the cost of deploying the sensor devices. True, a knowledgeable hacker may be able to break into the system, but it would likely be just as easy to throw hand grenades at the crane operator.

Wi-Fi is a relative newcomer in the embedded industry (if you don't count cellular phones and other consumer electronics. Industrial embedded wireless is pretty new, at least for inexpensive industrial control), the utility of wireless communications for embedded devices has long been known, but the technology and infrastructure just have not made it a good choice until recently. Since it is new to the embedded world, there are still many things to work out with securing embedded wireless devices. One potential issue is the excessive noise (radio noise, that is) that could be generated by hundreds or thousands of tiny wireless devices occupying a small area (such as a house or office building). With all the simultaneous connections opening and closing, and data flying everywhere, we need to be sure that these devices behave themselves—and that means they need to be secure.

One protocol poised to take over the embedded world is ZigBee, based on the IEEE 802.15 standard. ZigBee is a wireless protocol that is tailor-made for embedded applications. The protocol is designed with power consumption (or rather, the lack of it) in mind, and it supports peer-to-peer and mesh networking, both useful in embedded applications (and not part of the basic 802.11 protocol). ZigBee is a hot technology (at least when this book was being written), and hundreds of new applications are beginning to appear. ZigBee is a very flexible protocol, allowing for little or no configuration of individual devices when establishing a network. The protocol itself provides a few basic mechanisms for preventing replay attacks and for doing some basic authentication, but it really relies on AES for security. Not having a complete security protocol like SSL, ZigBee by itself may not provide all the required security for some applications, but for most implementations it is probably enough (and you could always implement SSL on top of ZigBee if you really needed it).

Wireless technologies are a hot topic in the embedded industry and likely will remain that way for a long time to come. We are still in the infancy of wireless communications, so there still is a lot of work to do. However, given the state of wireless technologies, their newcomer-technology status, and the buzz these technologies are producing, we dedicate the entirety of Chapter 6 to wireless technologies. In Chapter 6 we will look at the tangled mess of security in wireless protocols, especially in 802.11 Wi-Fi. We will also cover other technologies such as ZigBee and Bluetooth, and briefly touch on cellular applications.

Wrap-Up: Network Communications

This chapter has been all about network communications and the inherent security (or lack thereof) built into each protocol. As we saw, there are many ways to look at communications and where attacks or errors may occur. This brief tour of the protocols also prepares the reader for the upcoming chapters where we look at security protocols and how we can implement them for resource-constrained environments. In Chapter 3, we will look at security protocols and algorithms, covering both specific cryptographic algorithms and the protocols that use them.

Security Protocols and Algorithms

Network and Internet security are dominated by security protocols. A veritable sea of cryptic abbreviations and acronyms can confuse and confound anyone new to the discipline of security. We will look at some specifics here, covering such protocols as SSL, SSH, and IPSEC. These protocols will be introduced here, and we will cover their histories, but SSL will be given much more attention later on (SSL is so important that it has its own dedicated chapter).

As you may already know, the Internet works on the same principles as your typical Local Area Network (LAN), just at a much larger scale. For this reason, we can look at network and Internet security as being essentially the same. Similar problems plague both small networks and the Internet. Some of these problems are due to uncontrollable factors, such as weather and the unpredictability of end-users. The problems may be concerted efforts from malicious individuals or organizations that have something to gain from stealing information or making use of the network difficult. However, both purposeful and unintended failures in the system compromise the security of your application.

So far, we have covered some existing network communications protocols that may be used in an embedded environment. Almost every communications protocol has some type of built-in security, but it is up to the application developer whether or not to use those options. This obviously leads to security problems, since lazy or hurried engineers are likely to skip robust security features in an effort to get the product to market. Though this could be a problem for the aforementioned protocols, we will now look at some of the various protocols designed explicitly for secure communications. Some of them, such as the Secure Sockets Layer (SSL) are simply enhancements to existing, less-secure protocols (such as TCP). Others are stand-alone protocols that perform a specific service for the application—such as in the case of the Secure Shell protocol, or SSH, which provides a secure form of remote console (similar to Telnet).

Before we get started, we should point out a general feature of most secure communications protocols. In order to be generally useful, a secure protocol should not be limited to only supporting connections with known systems, but should allow for anonymous yet secure communications with *any* system. The reason for this is that it is extremely impractical to set up pre-defined trusted networks for some applications. Sometimes this is necessary, as when setting up a Virtual Private Network (VPN), but most applications generally use some type of authentication, such as a password, to trust the remote machine. For this reason, most dedicated protocols utilize some form of public-key encryption, which allows for the sharing of cryptographic keys with unknown systems. The problem with using public-key encryption, as was mentioned in Chapter 1, is that the algorithms are extremely slow— sometimes thousands of times slower than their simpler, symmetric counterparts. In order to deal with this problem, many of the protocols utilizing public-key cryptography use a hybrid approach that uses the public-key algorithm to securely exchange keys for a faster symmetric algorithm.

For now, we will take a quick tour of some dedicated security protocols to introduce the reader to the technology and learn of the security options available to application designers. The protocols introduced here will be covered in more depth later, as we pick apart each one and see what we can keep and what we can scrap to make it fit into a tiny memory. Just remember—DO NOT try to optimize the algorithms—but instead, look for extra features that can be removed without damaging the integrity of the security functionality you need. Some protocols, such as SSL/TLS, are so flexible that there are many options available to us—in the case of SSL, there are so many options that we dedicate an entire chapter to it (not to mention the fact that SSL is probably the single most important security protocol in existence)!

Protocol Madness

The IT industry and the Internet are dominated by protocols. As was seen in Chapter 2, there are many different protocols that play various roles in allowing two devices to communicate via a network. From the lowest levels of the hardware layer, to high level abstractions, protocols are everywhere. Logically, it follows that security is a concern for all these protocols, and even the security itself is defined by other protocols. In all of this, it can be difficult to know what is necessary for an application. Does the communications protocol I am using have sufficient built-in security for my application, or should I look for something better? In order to find your way in the complexities of network security protocols, we will introduce some actual security protocols and identify what they can and should be used for, how they work, and how we can adapt them for resource-constrained systems. As we mentioned, the Secure Sockets Layer (SSL) is so important that we dedicate an entire chapter to it later. For now, we will concentrate on more specific protocols that can be very valuable since they provide only a few services (some only one), and this allows us to use specialized (and therefore smaller and sometimes faster) mechanisms to secure our applications without having to resort to a more cumbersome general-purpose security protocol. Before we get into specific protocols, however, we are going to look at the security protocol and what it means, in general, for your application.

Standardizing Security—A Brief History

Security has long been a concern for computer applications developers, as there have been hackers wanting to compromise systems from the earliest days of computer networks. Indeed, the first networks were used by corporations and the military, both targets for espionage and both have information they want kept secret. In the beginning, security, like the machines being protected, was crude and often fatally flawed in execution. Not until there were those with the motives and cunning to push the boundaries of what well-behaved machines were supposed to do was there any great desire to mathematically ensure that a system was safe. However, with the growing threat of hackers in the 1970s it quickly became apparent that computer security needed some rigorous treatment if anything was ever to be secure at all.

There was a lot of work in the 1960s on security, but it was mostly unpublished. In the 1970s academics began to collect information on security and began analyzing the mathematical foundations of those ideas. They found various holes in the assumed security of preexisting systems and began a rigorous mathematical discipline to produce algorithms that would provide security and stand up to mathematical analysis. They abandoned the existing idea of "security by obfuscation," which is essentially the practice relying on the hiding of security algorithms from attackers. This turns out to be a very insecure practice, as there are many ways to attack the algorithms used for security without actually knowing the algorithm. That being said, keeping the algorithm secret can help with security—at least on a superficial level. It is now generally assumed by security experts that allowing rigorous public scrutiny of algorithms is better than keeping the algorithms secret, since there is a higher chance of finding problems in the algorithms before an attacker finds them and uses them for malicious purposes.

Besides abandoning traditional security ideas, another major development in the 1970s was the development of a new type of cryptography. Since ancient times, there had been only one basic type of security—the shared-key, or symmetric, cryptography where both parties must have knowledge of a secret key shared ahead of the communications. This was OK for some applications, such as military communications, but it presented some problems with computer networks, since there was no effective way to share the key without compromising security or making the whole network idea much less useful (if you can just hand the key to someone, why not just hand them the data?). Some pioneers thought about how they could protect data without having to share a secret ahead of time. The basic idea was to develop algorithms that have a two-part key, with one half being public to share, and the other half being secret to provide the security. Public-key security was thus born.

There were a couple of different public-key algorithms developed, but one was basically universally adopted due to a couple of unique properties. RSA, named for its inventors

Rivest, Shamir, and Adelman, was reversible, unlike the other primary algorithm of the day, Diffie-Hellman (also named for its inventors). This symmetry allowed for two different but complimentary applications. RSA could be used for basic public-key cryptography, using the public-key to send an encrypted message to the owner of its matching private key. The advantage was that with RSA, the private key allowed the owner to encrypt a message that could then be decrypted with the public-key, thus proving that the message was sent from the owner of the private key that matched the public-key. This allowed for authentication over a network—if Alice knows Bob's public-key belongs to Bob, then she can decrypt an identifier sent by Bob and she will know that only he could have sent it, authenticating its validity (assuming that Bob's key is still secret). With RSA and other public-key algorithms, the fledgling Internet finally had a mechanism that could make network communications safe and efficient.

As the Internet gained more users in the 1980s, primarily corporations and academic institutions, many different protocols were developed, many competing for adoption. However, since the Internet was used not by the public, but by large organizations, there were few standards developed—each organization just used their own security. Services provided by the network, like email and file transfer, just were not secured because either no one saw the need, or the performance penalty was unacceptable.

It was not until the early 1990s that security really began to gain a mainstream foothold. The invention of the World Wide Web led to an explosion of users, and it became more and more evident that security was needed badly. Companies like Microsoft and Netscape (you may recall that Netscape Navigator and Microsoft Internet Explorer became the two predominant web browsers in the mid 1990s) tried to increase the use of their products through the introduction of generic protocols that could be used to secure any and all Web traffic. In the end, it was Netscape who won, with their implementation of the Secure Sockets Layer protocol, or SSL. SSL caught on quickly, and finally settled at version 3, which is still in widespread use today (version 1 was never released to the public and version 2 was found to be fatally flawed and nearly completely insecure). SSL was finally standardized by the Internet Engineering Task Force (IETF), one of the standards organizations for the Internet, as Transport Layer Security, or TLS. TLS was introduced in the late 1990s and is available in essentially all Web browsers and servers. The existence of SSL was one of the major factors that made the e-commerce explosion happen.

Today, security is needed more than ever. More and more, everyday items are being networked. Ordinary items are finding new uses through network interfacing, and this trend will only continue to grow. With everything connected to the global Internet, it is obvious we need to protect all that information. Fortunately, there are ways to adapt security mechanisms and protocols to resource-constrained systems so that even the most modest devices can have some modicum of security. The rest of this chapter will focus on security protocols and mechanisms, and how we can adapt them to work in resource-constrained environments.

Standardized Security in Practice

As a result of the efforts of researchers and investment in the infrastructure, we today have a large number of standard algorithms, protocols, and models to choose from when securing an application. Indeed, as the Internet is dominated by standards, so too is the realm of cryptography and security. Some of these standards can easily be adapted to fit into a resource-constrained environment, while others suffer from system requirements that make it nearly impossible to make them work. We will now take a look at some of the most commonly used protocols and algorithms, and note which are best suited for an embedded environment, and which are not. When looking at these standards, remember that we need protocols and algorithms that are flexible and can work with limited resources, as well as remain effective for possibly extended periods of time. An embedded application will often last years longer than equivalent PC hardware, and the security needs to be at least somewhat effective into the foreseeable future. Obviously with the pace of technological innovation today, likely all of these protocols and algorithms will be obsolete within the next decade or so, but you do not want to choose anything that is on its way to being replaced or removed.

Cryptographic Algorithms

The state of security standards today is pretty exciting. New algorithms are being developed, and even new types of encryption are being created, like elliptic-curve encryption and quantum encryption. There are many options to choose from and it can be difficult to choose a particular algorithm or protocol.

Several venerable algorithms are being made obsolete, such as the Data Encryption Standard (DES), which was developed in the 1970s but has recently been showing its age. Originally, DES used a 54-bit key, and at the time, it was thought that 54 bits was large enough to not be broken using a brute-force attack in any reasonable amount of time. However, the advances in computing power in the 1980s were not predicted,[1] and by the 1990s 54-bit DES was broken using a distributed application. Today, a 54-bit key is easily broken. One answer to the aging DES was a variant called 3-DES ("triple-DES"), where the key actually consists of 3 54-bit DES keys strung together for a total effective key size of 162 bits. This provides a fairly effective level of encryption, but the general algorithm is

---

[1] Moore's law predicted in the 1960s that the density and performance of integrated circuits would double roughly every 18 months or so, but apparently none of the DES designers thought the prediction would hold true for so long.

slow, and other algorithms were needed for higher performance. In older applications where speed is not as much of an issue, 3DES is still widely used.

A lot of cryptographic algorithms were developed in the 1990s, many of which were compiled into Bruce Schneier's classic *Applied Cryptography*. One algorithm that has remained relatively popular due to its blazing performance is Rivest Cipher 4, or RC4 (Rivest is the "R" in RSA). Originally a trade secret of RSA Security (the company founded by the creators of RSA), RC4 was leaked to the Internet and has since become a very popular cipher, since it is extremely easy to implement, and it is exceedingly fast. However, RC4 has not been through a thorough mathematical analysis (as DES was) and it is a stream cipher (which means there are some caveats in using it—when we discuss WEP in Chapter 6 we will look at this in more detail), so though it seems to be secure, there is some doubt that accompanies the algorithm. These doubts have not prevented RC4 from being built into virtually every web browser as part of the SSL protocol, and a large number of secure Internet transactions are performed using 128-bit RC4.

With DES becoming obsolete and algorithms like RC4 not having the rigorous treatment to give everyone confidence, there was a big push to create a new standard algorithm. To create interest in developing a new standard, the National Institute of Standards and Technology (NIST) created a competition in the late 1990s that called for submissions of algorithms that would all undergo thorough public and internal analysis, with a committee to choose the final algorithm. A number of entries from top cryptographers were considered for what would become the Advanced Encryption Standard, or AES. After much public debate and scrutiny, the NIST committee chose an algorithm called Rijndael, named for its creators (the name was actually a combination of the author's names). Now commonly known as AES, Rijndael is the standard encryption method used by the United States government and numerous other organizations. Essentially every security protocol today supports AES (except for those created before AES—and those protocols likely use a variant of DES).

In contrast to the variety of algorithms described above, which are all symmetric-key algorithms, there are relatively few public-key algorithms in use today. This is primarily due to the dominance of one particular public-key algorithm, RSA. Named for its inventors (Ron Rivest, Adi Shamir, and Leonard Adleman), RSA is so simple to implement and so elegant in its execution that most other contenders are just not attractive enough to bother using. RSA was a patented algorithm, and for a number of years was available only under license from RSA Security, the company created by Rivest, Shamir, and Adleman to govern not only RSA but a number of other security algorithms, protocols, and services. However, the patent for RSA recently expired, and with no barriers remaining it has become the primary public-key algorithm in use today. Despite RSA's dominance of the field, however, there are some other algorithms that are available, such as Diffie-Hellman (one of the first public-key algorithms), PGP (Pretty Good Privacy, used for email encryption, among other things), and DSS (Digital Signature Standard, an open protocol that was designed for digital signatures). Due to the slow performance of public-key operations, all these algorithms are usually relegated to specific uses where the public-key encryption properties are explicitly needed, otherwise a faster symmetric algorithm is usually chosen.

Cryptography in Practice

Now let's take a look at a couple of algorithms a little more in depth. Due to their wide-spread use, we will focus on four common algorithms, DES/3DES, RC4, AES, and RSA. Please note that the descriptions of these algorithms in the following sections are derived from publicly available specifications and from Bruce Schneier's excellent reference, *Applied Cryptography*. Later on, in the case studies, we will look at the implementation of these algorithms and how we can apply them in our own applications. For now, we will just take a look at the algorithms, where they come from, why they are popular, and what they can and should be used for.

DES

The Data Encryption Standard was developed in the early 1970s by a team from IBM with some help from the National Security Association (NSA). Many of the details of the implementation remained secret for many years, as the NSA did not want to reveal what techniques it had up its sleeves. The algorithm IBM had created was originally called Lucifer and was modified and tweaked by the IBM team and the NSA as a response to a call from the NIST precursor for a government standard encryption scheme. The public response was enthusiastic but underwhelming (no one was really studying cryptography at the time), so the design from the IBM team was a shoe-in. DES was originally intended to be implemented in hardware and was designed around the technology of the time. However, the published specification contained enough information for software implementations to be designed once general-purpose computers became more accessible. DES by itself is a fairly quick algorithm, but its standard 56-bit key is too small to be practical anymore (a 56-bit key can be broken using brute force methods over a distributed network in very little time). Triple-DES, or 3DES, uses 3 56-bit keys to achieve a higher level of security, and is more standard today. Despite its age and suspicious involvement of the NSA in its design, DES has stood the test of time and still remains a viable cryptographic algorithm.

The implementation of DES has some small challenges in software, as a couple of opera-tions are trivial to implement in hardware but require some additional work to implement programmatically. The math behind DES is pretty heavy-duty (you can read all about it in *Applied Cryptography*), but the implementation is pretty straightforward, since most of the tricky math is actually wrapped up in a collection of constants that are used essentially to randomize the data being encrypted. These constants are put into what are called *substitution boxes*, or s-boxes, and *permutation boxes* (p-boxes). The real security of DES resides in the s-boxes and in the key—we aren't going to worry about why DES is secure (plenty of books and papers have been written on the subject)—for our purposes we can just assume that it is. We will instead look at the implementation and the tradeoffs we can exploit to make DES work for an embedded application.

One of the advantages of DES is that it was designed in a time when all hardware resources were scarce. The algorithm predates 32-bit and even 16-bit buses and is specifically designed to function in an 8-bit world. This property dates the algorithm, but it is also a boon to modern embedded developers, since the operations are designed for 8-bit computing. DES is also designed to be implemented directly in hardware, which can be a big advantage for performance. Conceivably, it would be very easy to add a DES coprocessor to an embedded design and not add too much to the cost.

Unfortunately, the use of DES is discouraged by pretty much everyone for new development and designs, so its utility is a bit limited. In fact, original DES should not be used at all, since it is considered too weak (see the explanation earlier in this chapter), so 3DES is really the only option. Despite the fact you shouldn't use DES anymore, it is still implemented in a wide array of applications, and if you are attempting to interface with a legacy product, it may require DES, so we include it for this reason.

RC4

RC4 is probably the simplest algorithm for securing information in existence. The entire operation is essentially an XOR and a swap in a 256-entry table. The operation is completely reversible (the same operation is used both to encrypt and decrypt), and extremely fast. RC4 was originally a trade secret of RSA Security, and was invented by Ron Rivest (the "R" in RSA). The introduction of RC4 to the public is rather unique as well. Someone managed to reverse-engineer the algorithm and this person then posted the source code on the Internet that was completely compatible with the proprietary implementation! Oddly enough, the public implementation was different enough from RSA's implementation that they could not claim it was stolen. Since the algorithm was not sanctioned by RSA or Rivest, the public version of RC4 was referred to as *Alleged* RC4 (ARC4) for many years. The algorithms are exactly the same though, and as a result, there is no effective difference between the two.

RC4's unique origins and its simple implementation have made it quite popular, and essentially every web browser and SSL implementation supports it. It can be implemented in just a few lines of C code, and is blazingly fast. The real downside to RC4 is that it has not been as rigorously studied as other algorithms, and this makes many security experts a little uncomfortable. Beyond that, the fact that RC4 is a stream cipher has some inherent drawbacks to implementation. Specifically, if an RC4 key is used twice to decrypt different messages, then the encrypted messages can be combined to retrieve the plaintext message without knowing the key. However, as long as you never reuse the same key (there are a lot of keys in 128 bits), then you should be OK (see below).

Other than the issues we mentioned, there are no known attacks on RC4 itself as of this writing, and it seems to be an OK cipher (the attacks we talk about are actually targeted at poor implementations of RC4 in a system). As an embedded designer, though, it is very hard to overlook RC4 for a secure application, since it is so small and fast. It is widely used, so compatibility is not an issue either. It is a good algorithm if you want a reasonable level of security for many applications, but be advised that the jury is still out, and there is always the possibility of a breakthrough that could make it obsolete. There are also a number of drawbacks to using stream ciphers that, if not accounted for, can lead to a compromise of the security.

One major caveat to point out with RC4 is that it is a stream cipher (as opposed to a block cipher like AES or DES). A major problem in using stream ciphers (or using block ciphers in a stream-cipher mode) is what is called a substitution attack. If the attacker knows the format and portions of the plaintext message, he can anticipate where parts of the message will be and actually modify the encrypted message without knowing the key. For example, it would be fairly trivial to change the word "Canada" to "Israel" in a message (note that the words are the same number of characters so the message would still make sense) if the attacker knew that the word "Canada" was in the message and knew its position within the message. This is especially problematic for banking or database transactions, since they may be automated and can be expected to follow a particular format. Protocols that utilize stream ciphers effectively use a checksum or hash of the message included in the encrypted payload to check for any tampering.

Another problem with stream ciphers is that you must never use the same key twice. The problem here is a mathematical property of stream ciphers—if an attacker has two different encrypted messages that use the exact same key then he can retrieve the message rather trivially. The solution to this problem is to use what is called an *initialization vector,* which is essentially a one-time random string of data that is combined with the secret key to produce a relatively unique key for transferring data (SSL and other protocols that use stream ciphers do this). The initialization vector must be large enough to ensure that the probability of using the same vector twice is extremely low.[2] This was one of the issues with WEP (Wired Equivalent Privacy, a broken Wi-Fi security protocol which we will

---

[2] The so-called "birthday paradox" illustrates (somewhat counterintuitively) that the probability of two arbitrary numbers in a random set being the same is much higher than pure randomness would seem to dictate. This is the basis of a birthday party trick, since if there are more than 20 people in a room, the probability of two people having the same birthday is close to 50% (you just can't pick which 2).

discuss more in Chapter 6), since it used sufficiently short initialization vectors to increase the probability of a match to a level easily broken by a computer. Roughly once in a few thousand connections there would be two connections that had the same initialization vector, allowing for the aforementioned attack to work.

RC4 (and other stream ciphers) can be used safely if the appropriate measures are taken to account for the caveats mentioned above. However, it is important to keep up on the state of cryptography to be sure no new attacks are found. If you really need security that is standard and backed by a lot of experts, you should look to AES instead—it isn't as simple or as fast as RC4 (it's close though, we port AES to a PIC later), but it is a standard used by the US government, so at least your application will be in the same boat as everyone else.

AES

As was mentioned above the Rijndael algorithm was chosen in the early 2000s by NIST to be the new government standard symmetric encryption algorithm, otherwise known as the Advanced Encryption Standard.[3] Intended as a replacement for DES, AES utilizes much of the knowledge gained in the 30 years between its development and the introduction of DES as a standard. AES is also designed to take advantage of modern computer architectures as well, so it is a little less cumbersome than its predecessor. The AES algorithm is a fairly simple design that lends itself to high performance and elegance of implementation. Also, one of the stipulations of the AES competition was that the algorithm should be designed to be easily implemented in hardware, so AES has some properties that make hardware acceleration a real possibility.

AES can be implemented completely in software or completely in hardware. More recently, a hybrid approach has been taken, moving some of the more expensive operations into hardware while leaving the rest of the implementation in software. This new approach has several distinct advantages. The hardware, not being dedicated specifically to AES, can be more general and less expensive. Having the hardware acceleration, however, makes the software smaller and faster. Several companies provide various levels of AES acceleration, even going so far as to add specific AES instructions into a processor so that no external hardware devices are needed. We will take a look at a chip that offers these instructions and an implementation that uses them when we look at the Rabbit 4000 microprocessor[4] in our case studies.

---

[3] AES actually defines various modes of operation, which makes it more of a protocol than an algorithm. However, the AES moniker is usually used to refer to the basic Rijndael algorithm, which is how we will refer to it throughout this book.

[4] DISCLAIMER: At the time of this writing, the author is employed by Rabbit Semiconductor, the maker of the Rabbit 4000.

Generally speaking, if you need a symmetric encryption algorithm for an application, you should use AES. It is the safest choice because it is the standard used by the US government and it is backed by some of the leading experts in cryptography. If AES is ever broken, it won't be your fault (and it wouldn't matter anyway, since everyone else would be in *big* trouble too).

RSA

The three algorithms we have looked at so far are all symmetric-key algorithms that are very useful if you have a good way to exchange keys with the person or machine with which you want to communicate. In some cases, a physical transfer of the key may be possible and appropriate, but more often, an electronic means of key delivery is more applicable. Today, in most cases, this method is RSA. Developed and patented by Ronald Rivest, Adi Shamir, and Leonard Adleman in 1978, RSA is the most well-known and probably most useful public-key algorithm. One of the most useful properties of RSA is that it can be used both for the basic public-key operation (I send you my public-key so you can encrypt a message to send back to me), and for authentication (I encrypt a message with my private key which you can verify came from me using my known public-key). This property makes RSA especially useful for protocols that utilize both a public-key operation and authentication, like SSL.

RSA is an elegantly simple algorithm with some extremely complex math behind it. Essentially, RSA consists of a function that utilizes some unique properties of large prime numbers and modular mathematics. The key generation for RSA involves selecting two very large prime numbers and multiplying them together. The trick is that if you know the prime factors used to generate the key then the RSA encryption function is simple to reverse (thereby decrypting an encrypted message). If you don't know those factors, then you have to find them by factoring a REALLY large number into its component prime factors, a process that takes an extremely long time using today's math and technology. It is possible that we may discover a fast way to factor large numbers in the future, and this would be a very big problem, since it would render RSA completely useless. This prime factoring can be considered similar to the brute-force search required for the naïve attack on symmetric algorithms. However, factoring a number is inherently easier than searching through all possible symmetric keys represented by a number of the same size. It is for this reason that symmetric keys are often 128 bits long, but a similarly-secure RSA key will be 1024 or 2048 bits long. Unfortunately, these large numbers are hard to deal with, even on a computer, so RSA is significantly slower than any of the symmetric algorithms we have looked at.

RSA is too slow to be generally useful, but since it does have specific useful properties, it is usually used along with a faster symmetric algorithm, usually to exchange symmetric keys.

For embedded systems, the performance of RSA is especially problematic, since an RSA operation on even a moderate-sized key (1024-bits is considered moderate as of right now) can take many *seconds* on slower processors. For some applications, this might be OK, since the system may not need super speedy response times. For other applications, such as web servers, this is definitely a problem, as the typical user will not wait 30–40 seconds for a webpage to even start loading. For this reason RSA should be used with some type of hardware acceleration if at all possible. Fortunately, like AES, there are a number of products available that either implement RSA entirely in hardware or provide some type of acceleration for software implementations. The primary part of RSA that benefits from acceleration is the modular math—literally millions of operations are done in a typical RSA operation and any improvement in their speed will result in a significant improvement in performance.

RSA is an extremely useful algorithm that is employed in thousands of applications. We will look more at how RSA is implemented, and what we can do to alleviate the performance bottleneck, when we look at SSL, and again when we look at the case studies later on.

Cryptography and Protocols

Most of the time, cryptographic algorithms are not used on their own, but rather as part of a complete security system or protocol. Indeed, as was mentioned earlier, RSA is pretty much always used with some symmetric algorithm as a key exchange mechanism. Security protocols are designed to address specific problems in communications. Many protocols are designed to be used for a particular application, such as the Secure Shell (SSH) protocol, which is designed to provide a remote text-based console, like Telnet but secure.* Some protocols are designed to be a general purpose solution, encrypting everything that is sent between two machines on a network. Examples of this type of protocol include SSL and IPSEC.

When choosing a protocol for an application, you have to look at not only the features that the protocol provides, but also how the protocol has proven itself in the field. SSL version 2 (the first publicly available version, implemented as part of the Netscape web browser) seemed to be secure, but was later shown to be fatally flawed. The replacement, SSL version 3, has been in use for almost a decade now, and seems to work pretty well. You also need to look at who designed the protocol—was it "design by committee" or were

---

* SSH is actually now a more general purpose collection of secure utilities, but its original intent was as a secure replacement for Telnet.

there some security experts and cryptographers involved? A recent example of why you need to research a protocol before using it is the case of the Wired-Equivalent Protocol (WEP), used by the Wi-Fi protocol suite to provide basic security for wireless transmissions. The protocol was designed by a committee that did not include the appropriate experts, and once the protocol went public, it did not take very long for some real experts to show that the protocol was fatally flawed. Having learned their lesson, the committee used some real experts and cryptographers to implement the replacement, called WPA.

Implementing a security protocol can be a bit tricky, since there are a lot of places where you can go wrong. Probably the best defense against improperly implementing a security protocol is to strictly follow good software engineering practices. You should also look into hiring a real security or cryptography expert as a consultant—an expert will know exactly where the weak points of an implementation are and help you to fix them. The other problem is that a security breach is unrelated to the protocol, residing in another part of the system entirely. This means you can easily fall into the trap of believing that your system is secure since you used a secure protocol, but neglecting the rest of the application can make all your efforts with the protocol meaningless. In the next chapter, we are going to look at some ways that we can implement a secure embedded application through the use of good programming practices.

When looking to a protocol for an embedded application, one property to look for is flexibility. An example we will look at later is SSL, a protocol that you can get away with only implementing essentially half of for a large savings in code. SSL also allows the implementer to choose what cryptographic algorithms to support. You can choose algorithms that are more suited to an embedded environment and leave out those that may not be as attractive. We will not go into the details of the flexibility of SSL right now, as Chapter 4 is dedicated to that particular protocol and how useful it can be.

The reason flexibility is important in a protocol is that security protocols (with good reason) are designed with security in mind first, and resources second. Obviously, a protocol that is extremely secure but takes forever to run will not ever be used, but most protocol designers today are working with nearly unlimited resources when compared to economy-priced embedded hardware. Their focus has been on the "big iron" market—PCs and mainframe communications where plenty of system resources are available. If we were to implement all the features of some protocols, the resulting code could be into the megabyte-plus range. For a target system with around 1 megabyte total space (code and data combined), this is obviously a problem. We need to be able to implement only what we need and nothing we don't—but we don't want to remove anything important. In later chapters, we will look at some ways to do exactly that by picking apart protocols and finding what it is that we really need.

Other Security Protocols

In Chapter 4 we will look specifically at SSL, by far the most widely used security protocol on the Internet. However, there are a lot of other secure protocols out there that may work for specific embedded applications. To finish out the chapter, we will look at some of the most popular alternative protocols and discuss their advantages and shortcomings for embedded development. Some of these protocols and algorithms are more specialized than the general-purpose SSL, which can be both an advantage and an issue. On one hand, the more specialized a protocol or standard is, the more likely it is that it can be implemented in a small footprint. On the other hand, the specialized nature of the protocols leaves less room for flexibility, but for some applications, this may not be an issue. The following is intended to be a brief overview of the protocols, rather than an in-depth analysis, so we can get to the discussion of the Secure Sockets Layer.

SSH

Sharing many similarities to a specialized version of SSL, the Secure Shell (SSH) protocol provides a secure Telnet-style console interface, as well as several other remote-access services. This is a highly useful tool for creating secure remote configuration interfaces, since it is geared toward remote shell functionality. Due to the similarity in abbreviations, SSH is often confused with SSL, but they are separate protocols developed by separate organizations (the Netscape Corporation created SSL, and SSH was developed by a Finnish researcher). The two protocols do share many similarities, such as the use of public-key encryption to exchange symmetric-key encryption keys and the inclusion of both authentication and encryption mechanisms. The current version of SSH is an updated variant of the original that provides more security and has been standardized by the IETF in RFC 4251.

The SSH protocol is an important security tool and as an implementation that is provided as a standard utility in many operating systems. Also, like SSL, there is an open-source implementation available (www.openssh.com). In this book, we will cover SSL in depth and leave SSH alone, since the challenges that apply to SSH are essentially the same as those found in implementing SSL for a resource-constrained system. For more information on SSH, check out RFC 4251 and OpenSSH.

IPSEC

IPSEC is a protocol designed to work at the IP layer, and is widely used for implementing Virtual Private Networks. IPSEC is a security framework similar to SSL, but with a bit larger scope. The IPSEC protocol, described in RFC 2401, consists of various security mechanisms and policies that are used to build different kinds of secure tunnels between devices. The protocol is quite broad, and implementing it for an embedded system would represent many challenges. We mention IPSEC because it is an important component of network security, but SSL (or TLS) can be used for creating similar secure networking tunnels. SSL has the advantage of being a simpler protocol that is built on top of the underlying network stack (usually between the application and a TCP layer), whereas IPSEC is implemented lower in the stack, at the IP layer as seen in Figure 1. For an embedded system, implementing IPSEC would require additions to the network stack that may or may not be possible, depending on the TCP/IP implementation used. For these reasons, the author has chosen to focus on SSL/TLS instead of IPSEC as the general security protocol example for resource constrained embedded systems.

Before we finish with IPSEC, it should be noted that it still could be used for an embedded application if the actual embedded devices were networked with an IPSEC-enabled router. In Figure 2 you can see how a network of embedded devices can be placed behind a single

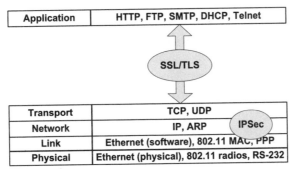

Figure 1: IPSEC vs. SSL in the Network Stack

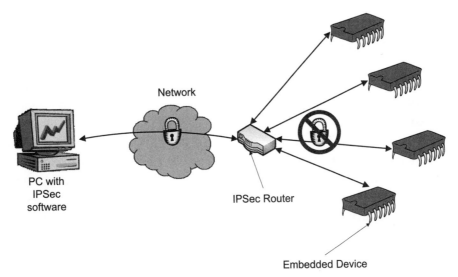

Figure 2: Embedded Devices Behind an IPSEC Router

IPSEC router, which is in turn connected to the public network. The router can setup a secure VPN tunnel over the public network, and the devices are not required to do anything to provide additional security. This assumes, of course, that the network of embedded devices is somehow physically isolated from eavesdroppers or attackers, but for some applications this may be a valid solution. For example, an application that monitors a remote warehouse could have a private network of sensor devices locked inside a building. That network could in turn be connected to the IPSEC gateway, which provides the security for all information travelling over the public Internet. This application would be secure, but does not really follow with the spirit of true embedded security, so we will not discuss this type of application further.

Other Protocols

Finally, to close out our discussion on security protocols before we jump into our in-depth discussion of SSL in the next chapter, we will briefly mention a few other protocols and standards. There are literally hundreds of protocols available for security, many of them freely available through the IETF RFC standards. There are also numerous security packages available from various vendors, and some of them are quite popular, but those products frequently use predefined and proven algorithms and mechanisms. For crypto-graphic algorithms, one need only look at the size of *Applied Cryptography* and glance through it to learn that those algorithms number in the thousands. In this chapter we have covered the most popular and recognizable standards, since they are proven in thousands of applications. There are many references on the more obscure algorithms, but they are gener-ally of more academic than practical interest. A few remaining protocols that bear mentioning here are: HTTP authentication, the HTTPS protocol (HTTP over SSL), secure FTP, and PGP.

HTTP authentication is basically a simple mechanism that provides a minimum level of integrity checking. The only thing HTTP authentication does is authenticate a username and associated password for an HTTP session, but it does not do any encryption or other protection of information. HTTP authentication can use either a plaintext mode or a digest mode. In the plaintext mode, the protocol sends the username and password as plaintext over the network and checks them against a database of stored usernames. Other than maybe keeping out nosy (and stupid) kids, there is really no use for this mode. In digest mode, the authentication is a little more secure, since it utilizes a challenge-response mechanism that requires the password to be hashed before being sent. This is slightly more useful than the plaintext mode, but it still provides only a thin layer of security that really is an illusion. It may keep your kids from modifying your personal webpage, but it really doesn't fit the bill for anything else.

Despite the basic uselessness of HTTP authentication on its own, it actually does have a use when it is paired with HTTPS. The HTTPS protocol, described in RFC 2818, is simply the

HTTP protocol as used over TLS (or SSL). There are a few slight differences in the way the protocol works, but those differences will not affect most applications. We will actually discuss HTTPS more in later chapters, especially when we look at SSL and client-server protocols, but we mention it here for its use with HTTP authentication. The SSL protocol provides a secure channel between two devices on a network, and it even provides some level of authentication via the use of digital certificates. However, the authentication may not be enough in some cases, as when an SSL server does not provide client certificate authentication (SSL dictates that clients perform authentication on server certificates, but the reverse operation of servers authenticating clients is optional). If you have implemented an embedded device with an HTTPS interface and it does not support client certificate authentication, then anyone can connect to the device. The channel will be secure, but the device has no way of authenticating the user on the other end, unless you implement something yourself. This type of situation is where HTTP authentication can be useful as seen in Figure 3. The authentication mechanism is already built into HTTP, so all you have to do is enable it for your HTTP interface. Once the SSL session is established, the HTTP server will perform password authentication as normal, but in this case, the username and password will not be sent as plaintext over the network. As a result, the authentication is secure. This exact setup is actually quite an advantage for embedded applications since the HTTP authentication mechanism is much easier to implement (if you have an HTTP implementation, you probably already have it) and lighter-weight than SSL authentication. When we talk about SSL in the next chapter, we will discuss how we can build a secure

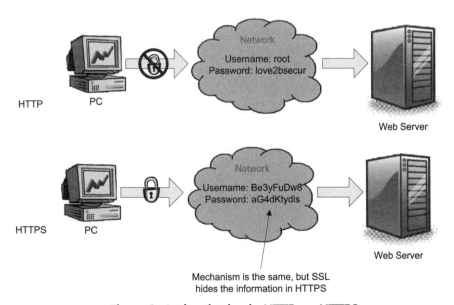

Figure 3: Authentication in HTTP vs. HTTPS

application in a small footprint by leaving out optional features such as the server-side authentication of clients.

The last two security technologies we will mention are secure FTP (FTP over SSL) and PGP.[5] Secure FTP (not to be confused with SFTP, the Simple File Transfer Protocol), is simply the implementation of FTP over an SSL connection. It will behave in the same manner as FTP, but the information being transmitted will be encrypted. Quite different from Secure FTP, PGP is an overloaded name—it refers to a product, a company, and a technology. Originally written in the early 1990s, PGP started out as a program to secure emails using public-key cryptography (it used its own algorithm). Still in use today, PGP is provided in commercial products that provide various levels of security and authentication. The commercial products are targeted at specific platforms, but the original technology could be adapted to an embedded application. This would, of course, require you to obtain the source code for the old PGP technology, which is likely protected by copyright, but it would be an interesting project to see if it could be ported to an embedded system— encrypted email could be a useful feature in many embedded applications.

Wrap-Up: One Protocol to Rule Them All . . .

We have only scratched the surface of security in this chapter, but it should be enough to get started with. This book is about learning what is out there and how it can be applied to embedded applications, so we have not gone into depth about the details of the algorithms we have discussed (there is plenty of material on that elsewhere). However, the next chapter covers a single protocol in depth, since it is so important to computer security in general. In the next chapter we will look at a protocol that has one of the most recognizable abbreviations in computer security, a protocol that is used in virtually every secure Web transaction today: the Secure Sockets Layer.

---

[5] PGP is a trademark of the PGP Corporation.

CHAPTER 4

The Secure Sockets Layer

The Secure Sockets Layer is so important to Internet security that we have devoted this entire chapter to it. SSL is the de facto standard for secure Internet transactions. It has achieved this status by being not only secure, but being highly generic and imminently practical as well. SSL exists in the network layer between TCP and your application, providing blanket security to all data transferred over the network. The API for an SSL implementation is typically very similar to the standard network sockets API (POSIX-style). For this reason, it is simple to transform any plain TCP/IP application into a secure Internet application with very little effort.

In this chapter we will look at how to implement SSL for embedded platforms. The standard is very flexible in algorithm choice and features, so it is easy to pick and choose the features from a requirement list. Most PC-based implementations of SSL are mono-lithic—all features are compiled into every application. However, due to the specialized nature of embedded applications, we will look at a more modular approach, allowing the developer to enable only the features that are needed, leaving the rest out of the final binary.

Aside from looking at structural procedures for making SSL work on embedded machines, we will also look at optimizations. Most of these will have been covered in previous chapters, since SSL uses common algorithms, but there are a few more tricks we can apply to make SSL work. We will look at hardware assistance as well, since SSL makes use of public-key cryptography, and some vendors have chosen to implement part or all of SSL in a hardware solution. We will also look at implementing SSL in hardware, as part of an FPGA, or as a stand-alone chip. Again, the advantage of hardware implementations is the ability to use larger keys and achieve the same or better performance. We will look at the tradeoffs between these hardware solutions, and a more flexible software implementation.

SSL History

The Secure Sockets layer had a modest beginning as a project at Netscape Inc. in the mid 1990s to secure Web transactions for their (then-dominant) Netscape Navigator web browser. SSL version 1 was an internal project that was never released. Version 2 was publicly released and quickly became a standard way of conducting secure web transactions, eventually beating out other secure offerings from other companies such as Microsoft. Unfortunately, some poor decisions in Netscape's implementation of SSL version 2 led a couple grad students to successfully break the encryption utilizing attacks on the random number seeds chosen (which included the system clock value—a number that is easily predicted and controlled). The attack revealed that an SSL-secured connection could be broken by a clever attacker in a manner of minutes. As a result of the compromise, SSL version 2 was deemed insecure and unfit for any real transactions. Despite this enormous shortcoming, SSL version 2 is still included in many popular web browsers, though it is now often turned off by default. You should never enable it, unless you want to try your own hand at breaking it.

In the aftermath of the SSL version 2 compromise, Netscape scrambled to provide a fixed implementation of SSL. SSL version 3 was born. At the time SSL version 3 was started, the Internet Engineering Task Force (IETF) began a project to implement a standard version of SSL that was designed to become the eventual standard. The standard, called Transport Layer Security, or TLS (a name generally disliked by the committee members but not hated as much as the other contenders), is now complete and described in RFC 2246. Due to the time it takes for a committee-designed standard to be completed, however, the Netscape version was introduced to the industry much faster than TLS, and version 3 soon became the dominant secure web protocol. TLS is structurally identical to SSL version 3, with a few minor enhancements to the way cryptography is handled, so it is sometimes referred to as SSL version 3.1. SSL version 3 to this day has not been compromised (at least as far as anyone can tell), but it is generally assumed that TLS is more secure and should be used for all new development. TLS is beginning to become the dominant standard, but SSL version 3 is still widely used and is still considered a practical way to secure web transactions. Due to the close similarity to TLS, SSL version 3 can be supported in a TLS implementation with a minimum of effort. From this point on, we will refer to TLS and SSL version 3 collectively as "SSL."

Pesky PKI

The SSL protocol provides blanket security for network communications by utilizing the advantages of both public-key cryptography and symmetric-key cryptography. As we talked about earlier, there are definite benefits to public-key algorithms because we don't have to figure out how to safely exchange keys with the person (or system) with which we wish to communicate. We can just send out our public-key and in receiving keys from others, we can communicate freely and securely. Well, that's not *entirely* true, since someone could potentially provide a public-key and lie about who they really are in order to entice you into communicating sensitive information to them. This exact problem is at the heart of how SSL is typically deployed in web browsers and servers. Before we delve into the deep technical aspects of SSL implementation, we will take a short detour and look at the authentication mechanism provided by SSL, and how this mechanism is used in practice to provide a notion of trust. However, keep in mind that the way SSL is currently used does not mean it is the only method of authentication that can be used—in fact, the mechanism by which SSL certificates are distributed has nothing to do with the SSL protocol. However, SSL is most frequently paired with the distribution mechanism described below, so we will spend a little time looking at it before talking in depth about SSL.

To solve the issue of authentication (namely, the secure distribution of SSL certificates), a few companies put their reputations on the line to provide the public with signing services. Companies such as Verisign provide these services then contract with browser and server makers to provide a "known" public-key in those applications. Assuming that "In Verisign we trust" can be taken as relatively true, then we have a relatively high confidence that any certificate we receive that has been signed by Verisign is valid and correct, and has been sent by the person or organization identified by said certificate.

The whole concept of Public-Key Infrastructures (or PKI) relies on the inherent trust in companies like Verisign to do a satisfactory amount of due-diligence on each certificate they sign. Assuming that Verisign is relatively trustworthy, then this is not a terrible system. In fact, it pretty much represents the backbone of e-commerce as we know it. Any time you browse to your favorite online stores, you are effectively saying that you trust Verisign and the browser and server makers with your financial well-being. This method by which SSL certificates are deployed makes some security experts nervous, but it has proven to be highly effective in promoting "secure" transactions on the Web.

SSL makes PKI possible through the construction of the digital certificates it uses for authentication. Since signing a digital certificate with a private RSA key, for example, is able to be chained any number of times back to a common root, one company can provide that root certificate as part of standard web application implementations and extend their trust to numerous third parties. SSL is designed to use any number of root certificates for authentication, so just a few companies provide root certificates, and everyone can use the system.

To bring this discussion back into focus for the purposes of embedded systems development, let's ask the obvious question: Just how many of these "root" certificates are there, and does my embedded application need to store *all* of them? It turns out that, like almost everything we have looked at in this book, it depends on the application itself. If your application is a webserver that will be deployed on the public Internet, then all you need to do is store a single certificate that you pay to have signed. If your device is designed to connect to any number of public Internet access points, then you will probably need to store at least a dozen or two root certificates, which unfortunately can be as large as a few kilobytes each. It is quite possible that a public SSL client would need as much as a megabyte of space to store all the relevant certificates to assure the highest probability of compatibility with any unknown server. However, it is likely that most embedded applications will either be an SSL server implementation, or they will not need to connect to every server ever created. For this reason, we will look at other methods for distributing SSL certificates that are more practical for embedded systems.

PKI Alternatives

As it turns out, the PKI system used for e-commerce on the Web is not the only way to deploy SSL certificates. A lucky thing that is, too, since many embedded applications would be hard-pressed to provide the space required to store all the root certificates. PKI is simply one implementation that has been pushed by large companies for web browsers, but we have the power to do things differently. In order to provide authentication for SSL in embedded applications, we will look at a couple of different options for deploying certificates that in many cases are more applicable and practical than PKI. In Figure 1 we can see the relative differences in the different deployments.

The simplest path to SSL authentication is to create your own self-signed certificate. This essentially means that you will create a certificate with the appropriate public-key/private-key pair, put the public-key in the certificate, and sign it using the private one. As it turns out, you don't have to use Verisign after all! Once you deploy your application, authentication is simply a matter of using the private key you created to authenticate the certificate being provided by your application. If your application is a web server, then you can install the certificate into the browser's root certificate cache—from that point forward, the browser will always accept that certificate (assuming it hasn't been tampered with or corrupted). Alternatively, if your application is an SSL client, then your server can provide a certificate signed using the same private key, which the application can then verify directly using the single certificate. For simple applications, a self-signed certificate is the only way to go.

So we can create a self-signed certificate, but what if we have to deploy hundreds of devices, each with a unique certificate? Well, you could pay anywhere from $100 to over $1000 to get each certificate signed by a public signing company or organization, or you can do it for free and be your own CA. How does *that* work? Remember that we can self-sign a certificate, so there is nothing keeping us from signing other certificates. All you need to do is to create a single root certificate (which simply means it is self-signed and will be used to sign other certificates), and then use that certificate's private key counterpart to sign all your other certificates. If you install your single root certificate in a web browser, then you have automatic authentication for all the certificates you create and sign using that certificate. As you might have guessed, this particular method of certificate distribution is

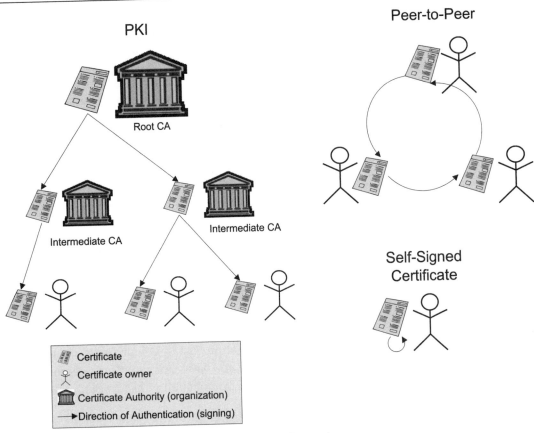

Figure 1: PKI Alternatives

quite useful in venues other than embedded systems, such as providing authentication on a corporate intranet, for example.

Creating a self-signed certificate and going into the business of being your own CA are both very reasonable ways to provide basic SSL authentication for your application. However, it may be that neither those methods nor PKI are suitable for what you are looking for. Fortunately, there are some other ways of distributing trust in the form of certificates, and most of them center on some concept of peer-to-peer networking.

There are various implementations of peer-to-peer authentication, and the idea hasn't caught on for e-commerce, so we will just stick to the basic ideas behind peer-to-peer authentication mechanisms. The basic idea is that of extending trust. As an example, start with someone you know and trust. Assuming that person is trustworthy, you can then assume that someone they say is trustworthy is probably trustworthy. This is fine unless that third party deceived your friend, but then again, the liar could deceive you as well. Back to

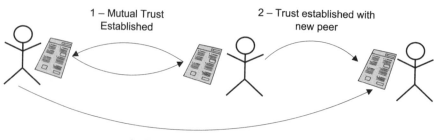

1 – Mutual Trust
Established

2 – Trust established with
new peer

3 – Trust Extended to new Peer

Figure 2: Peer-to-peer PKI

the point, though—if you can extend your "network of trust" by using mutual friends to verify new people that enter the network, you can extend trust to just about anyone. Figure 2 shows a simple peer-to-peer PKI example.

SSL Under the Hood

Now that we have covered the basics of authentication in SSL, it is time to dig in and look at what makes SSL tick. To start off, we will look at the basic structure of the communications layer of the protocol, starting with the backward-compatibility handshake.

One of the issues with SSL is keeping backward-compatibility with older versions. After all, it would be better to communicate using SSL version 2 than nothing at all (well, not much better, if your information had any value, but you get the point). The other action that can occur is that two SSL implementations may actually both speak newer versions of SSL or TLS, and they can negotiate up to the most recent version they both support. This backward-compatibility feature can be optional, allowing for newer implementations to communicate directly but the feature is required at the very beginning of the SSL handshake, so we will look at it first.

The SSL handshake begins with a network connection of some variety, in most cases the connection is a simple TCP/IP network connection (the lower-level and physical layers do not matter). SSL is essentially a TCP/IP application, and for the most part does not require any knowledge of the underlying network mechanism. That being said, SSL does rely on the network protocol being used to handle dropped packets, so a reliable protocol like TCP will work just fine for the lower level, but an unreliable protocol like UDP will not. This requirement of SSL does not reduce the security of the protocol or cause any other problems, even if the TCP connection is compromised. SSL will be able to detect any attempt to corrupt the information using the message digest authentication mechanism (discussed later in this chapter), and the encryption prevents eavesdropping. The problem

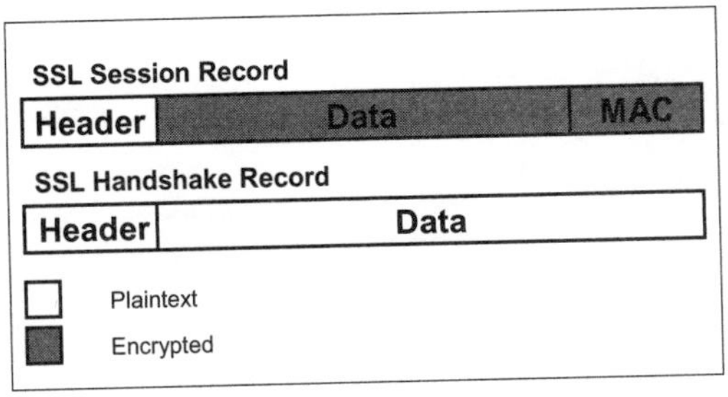

Figure 3: SSL Record

with using an unreliable protocol under SSL is that SSL will flag every dropped packet as a potential attack.

Once the network connection has been established, the SSL handshake itself can begin. The SSL communication layer is similar to other network protocols in that it consists of a header containing protocol information and a body that contains the information being transmitted. In SSL, these units are called *records* (see Figure 3) and can be thought of as being analogous to the TCP frame and IP packet. In the handshake, the SSL records are unencrypted, since the encrypted connection, called the SSL *session*, has not yet been established. The backward-compatibility handshake begins with an unencrypted SSL record called the *Client Hello*, which is sent by the SSL client wishing to establish the SSL session with a particular server. By convention, the application initializing an SSL connection is always the client, and follows a specific set of actions that correspond to actions on the SSL server.

The Client Hello message kicks off the whole thing, as we see in Figure 4. In backward-compatibility mode, the Client Hello is actually an SSL version 2 message in an SSL version 2 record, which differs slightly from the record in later versions. Originally, SSL used some fairly simplistic methods for generating the encryption keys and verifying that the handshake was not tampered with. As a result of the simplistic nature of these methods, SSL version 2 was compromised as we mentioned earlier. In SSL version 3 and TLS, the methods were improved, resulting in a more complex handshake protocol. With the backward-compatibility mode, the initial version 2 Client Hello message is translated into a version 3 (or TLS) Client Hello, and the handshake proceeds as usual. If the client truly supports only SSL version 2, then the connection will be aborted when the server sends a Server Hello back to the client indicating that it supports only SSL version 3 or TLS.

The Server Hello is a message that is built by the server upon receiving a Client Hello. The Client Hello message contains a number of *ciphersuites* that the client supports and is willing to use. A ciphersuite is simply a collection of algorithms and some specifications for the algorithms represented by an enumeration. For SSL, the ciphersuites consist of a public-key algorithm, such as RSA, a symmetric-key algorithm such as AES (also called the *bulk*

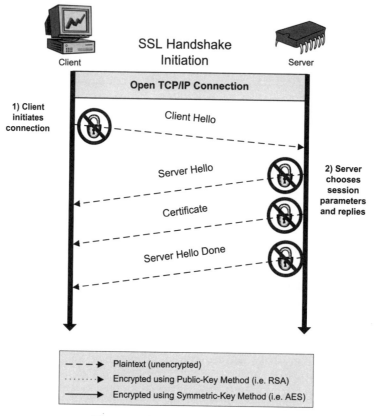

Figure 4: SSL Handshake Part 1

cipher), and a hashing algorithm. The Client Hello also contains a string of random data that is used as a seed for generating the *session keys* later (we will explain what session keys are and how they are generated a bit later when we get to the key-generation phase of the SSL handshake).

The server selects the ciphersuite it wants to use based upon its capabilities and the priorities set by the application designer or user for selecting cryptographic algorithms. This is the negotiation phase of the handshake—the ciphers actually used for the communication are determined by a combination of what the client offers and what the server can support. The server stores away the information from the client (such as the random data used later for seeding the session key generation) and generates a Server Hello message, which is sent back to the client. The Server Hello message contains the ciphersuite selection, a server-generated string of random data used in combination with the client random data for seeding the key generation, and some additional information to get the session started. Another thing that both the client and server do in this initial phase is to begin a hash that will eventually represent all of the messages in the entire handshake. When we get to the

Finished messages at the end of the handshake, we will see how these hashes are used to determine the integrity of the handshake process.

Following the Server Hello message, the server sends a message containing its digital certificate. The SSL certificate contains authentication information about the server that the client can use to verify the authenticity of the server, as well as additional information about the server and, most importantly, the certificate contains the public-key that will be used to exchange the information used to generate the session keys used by the secure SSL tunnel once the handshake is complete and the session is established. The certificate also contains a digital signature (an encrypted hash) that the client can use to verify the information in the certificate. The digital signature is usually signed by a third-party Certificate Authority (CA) using the reverse-RSA operation (encrypting using the private key, decrypting using the public-key). The client stores a number of digital certificates from different CA entities, and uses the public-keys stored therein to decrypt incoming server certificate signatures. In this manner, a client can verify an unknown server through the concept of distributed trust. If the client trusts the CA, and the CA trusts the server, then the client can trust the server as well. As long as the CA is reputable and checks up on all the servers it signs certificates for, this system works pretty well.

SSL is designed so that the server can send any number of certificates to the client, the idea being that the client should be able to use one of the certificates to authenticate the server. For this reason, the server Certificate message can be duplicated any number of times, once for each certificate the server sends to the client. In order for the client to know that the server is done sending certificates, the server sends a final message called the "Server Hello Done" message, indicating to the client that all certificates have been sent and the next phase of the handshake can begin.

The second phase of the SSL handshake begins when the server has sent the Server Hello Done message and the client begins the key exchange phase, as seen in Figure 5. The first thing the client does is to parse the server's digital certificate and verify the server's authenticity. It does this by first decrypting the digital signature using a locally-stored public-key (from a CA certificate), and comparing the decrypted hash to its own hash of the server certificate's contents. If the hashes check out, the client then proceeds to check the certificates *Common Name* with the address (URL or IP address) of the server for additional verification. It also checks a field in the certificate that indicates the effective date range that the certificate is valid for. If the address does not match the Common Name, or if the certificate has expired, the SSL connection is dropped. Optionally, and what is usually done in web browsers supporting HTTPS, the client can opt to continue the handshake even if one or all of the authentication checks fails. This usually results in a dialog box warning the user of the browser that something is amiss. If the user chooses to ignore the warnings by selecting to proceed anyway, or if the checks are all successful, the handshake proceeds.

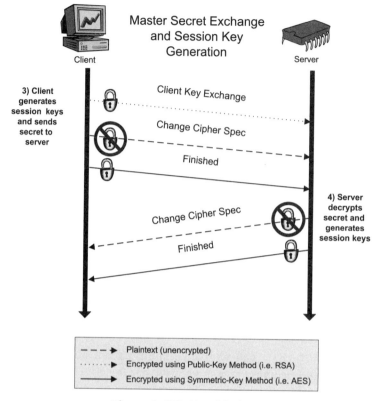

Figure 5: SSL Handshake 2

The client uses the information from the client and server Hello messages to generate the session keys, which will be used with the bulk cipher to securely send data back and forth once the handshake is complete. The client generates a random string, called the Pre-Master Secret, which is then encrypted using the server's public-key (obtained from the server's digital certificate). The encrypted Pre-Master Secret is then sent to the server in the Client Key Exchange message. Upon receiving the message, the server can decrypt the data using its private key (the complimentary half of the public-key sent in the certificate earlier). Following the Client Key Exchange message, the client sends the last message in the handshake that is not encrypted using the session keys (which we will talk about in a moment). This message is called the Change Cipher Spec message, and it indicates to the server that all following messages will be encrypted using the session keys. The final handshake message sent by the client is called the Finished message, which is encrypted using the session keys. In a bit, we will talk about what the Finished message entails and how the server ends the handshake, but first, where do these session keys come from?

Looking back at the beginning of the handshake, we mentioned that the client and server both generate a random string of data that is communicated in the initial "hello" messages.

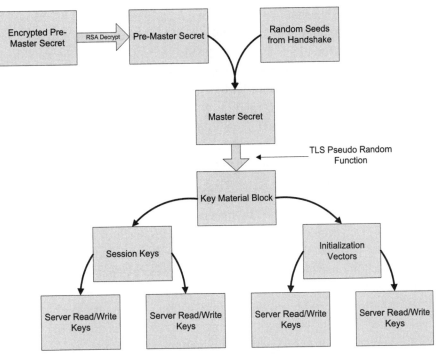

Figure 6: Key Material Block from Master Secret

We also said that the client generates a string of random data called the Pre-Master Secret that is used for generating session keys later. The Pre-Master Secret is small enough to be easily encrypted with the public-key algorithm being employed in the handshake, but large enough that it is not easily broken (64 bytes is a perfect size for a single 512-bit RSA-encrypted block, for example). The Pre-Master Secret and the random data sent during the initial handshake messages are used to construct the Master Secret on both the client and server—it is generated on the client as soon as the Pre-Master Secret is created, and on the server after the public-key decryption is performed (thus revealing the Pre-Master secret to the server). The Master Secret is essentially a cryptographically secure hash of the other data that is always 48 bytes in length. The Master Secret is used to generate what is called the *key material* block, essentially an arbitrary-length expansion of the Master Secret from which the session keys are derived as seen in Figure 6.

Generation of the key material block is done using a hash expansion algorithm that differs slightly between SSL version 3 and TLS (we will not cover what SSL version 2 does). In SSL version 3, the hash expansion is defined as part of the protocol. TLS uses a method called P_hash, which is composed of 2 hash algorithms, MD5 and SHA-1, and a hash wrapper called HMAC. HMAC, described in RFC 2104, provides a mechanism to add keying to hash functions. Essentially, both the SSL version 3 and TLS mechanisms are the same—the SSL version 3 mechanism is based on an early draft of HMAC. Both mechanisms do the same thing, which is to take the Master Secret and expand it into the arbitrary-length key material block. For typical SSL implementations, the key block is a little over 100 bytes, which requires 5 or 6 iterations of the key generation mechanism, the output of SHA-1 is 20 bytes (even with HMAC), so the algorithm strings several output strings together. The keys for HMAC (and the SSL version 3 mechanism) are part of the initial data from the Master Secret and serve to provide an additional level of security. Finally, the results of the expansion (using P_hash in TLS) are run through a final process that in TLS is called the *pseudo-random function*, or PRF. The result of the PRF operation is the final key material block.

The key material block represents the raw material of the session keys. Once the block is generated, the SSL/TLS protocol divides it up into pieces, each of which is used for a session key or other purpose. A general SSL session using a stream cipher will extract 2 MAC secrets, which are used to seed the integrity Message Authentication Codes (discussed below), and 2 session keys, one each for reading and writing. If a block cipher is used, the key block is further partitioned to produce 2 initialization vectors for the block cipher to use. You will notice that SSL generates 2 of everything, which is additional security precaution. It assumes that if one direction of communication (i.e. client to server) is broken by breaking the key, the other direction will still be relatively safe since it uses a separate key entirely. The same thought goes into the MAC secrets and the block cipher initialization vectors.

Now we go back to the handshake. The client has generated the keys and MAC secrets by the time the Pre-Master Secret is sent, so that it can immediately send its Change Cipher Spec message and begin the encrypted session by sending its Finished message to the server (the first message that is actually encrypted using the generated session keys). Upon receiving the public-key encrypted message from the client, the server generates *the same key material and session keys from the Pre-Master Secret as the client* as seen in Figure 7. This last point is key (pun intended)—even though the relatively small Pre-Master Secret is the only "secret" data shared by the server and client, they utilize identical methods to derive the same exact key material block, which is what allows them to communicate later. Once the server is done generating keys, it sends its own Change Cipher Spec message back to the client, indicating that it has generated its own session keys and is ready to initiate the secure session.

Finally, to close out the handshake process, the Finished messages are sent, one from the client, one from the server, as seen in Figure 8. The Finished message for both the client and server contains a hash of all the previous handshake messages (note that the server hashes the client's Finished message for inclusion in its own Finished hash). The message is the first to be encrypted using the newly generated session keys to avoid tampering. Both

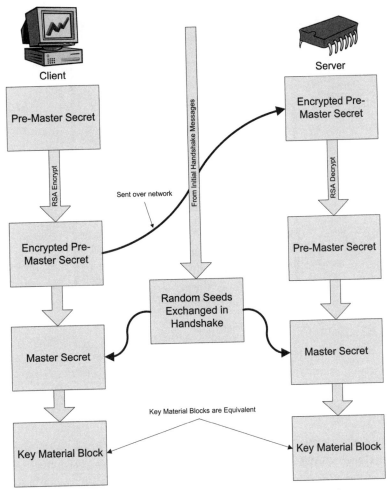

Figure 7: Simultaneous Generation of Keys on Client and Server

the client and server have a local hash of the handshake messages and the hash received in the other's Finished messages. Comparing the local hash to the remote hash (after the messages are decrypted) provides a final check that the handshake was not corrupted or tampered with in any way. Assuming the hashes check out, at long last the encrypted session begins.

The SSL Session

The handshake makes up the bulk of the SSL protocol, since it is the basis for all the security of the session. The session itself is fairly boring, since all that really happens is an encryption using the session keys derived during the handshake. One difference between a

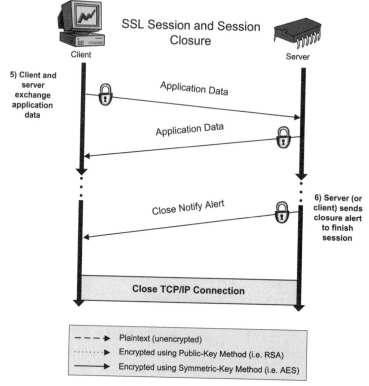

Figure 8: SSL Handshake 3

"regular" encrypted message and an SSL message (the SSL Record) is the use of a Message Authentication Code, or MAC, for verifying the integrity of messages. The MAC is a simple hash using an algorithm determined by the ciphersuite being used—a couple examples of hashes used are SHA-1 and MD5. The SSL MAC is actually slightly more than a hash of the data being sent, since it is actually seeded using the MAC secrets derived along with the session keys. The purpose of the MAC is simply to verify that the data in the SSL record was not compromised or tampered with, both things an attacker could do to disrupt communications without needing to break the actual encryption.

The session continues as long as both sides are willing to talk. At the end of the session, either the client or the server sends an *alert* to the other side to confirm that it is finished communicating. An SSL alert is simply an SSL record containing an encrypted message (again using the session keys) that usually indicates something abnormal has happened. In the case of the final alert, called close_notify, it is actually expected behavior and in fact it serves an important purpose. Without the final alert to confirm that communications are to be halted, an attacker could perform a *truncation attack* as seen in Figure 9, which basically means that not all of the data gets to its destination. This may not immediately seem like a

Attacker severs connection or intercepts packet, then
tricks sender into believing the packet was received

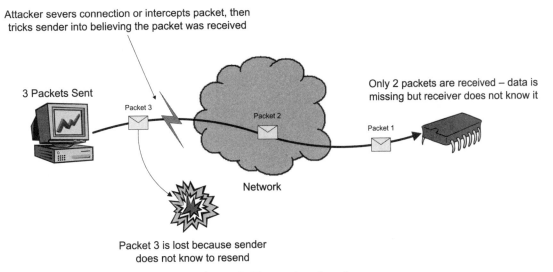

3 Packets Sent

Packet 3

Packet 2

Packet 1

Only 2 packets are received – data is
missing but receiver does not know it

Network

Packet 3 is lost because sender
does not know to resend

Figure 9: Truncation Attack

major issue, but think of bank transactions—you probably wouldn't be too happy if the
last few deposits you made simply disappeared since no one knew to check for them. The
close_notify alert allows both the server and client to verify that all information has safely
made it to its destination.

SSL in Practice

SSL is a fantastically successful protocol. Implemented in practically every web browser
and web server, it provides the security backbone for the majority of Internet transactions.
As the latest incarnation of SSL, TLS is being used even in other protocols as a
fundamental building block for security in wireless and mobile applications. One reason for
the success of SSL is the fact that it has an answer for just about every attack out there and
provides multiple layers of redundant security, making it a robust and provably safe protocol. Another reason SSL is so widely implemented is that the protocol is flexible—it defines
how cryptography should work together to provide security without actually defining what
"cryptography" should be. Even the key generation, which uses standard hash algorithms,
can be changed if needed. This flexibility is what makes SSL a prime candidate for implementation as a general security protocol for practically any embedded system. In each of the
case studies at the end of this book, SSL features heavily. The PIC application does not use
SSL directly, but the command protocol that is developed for use with the AES algorithm
derives some ideas from SSL, most notably the use of checksums and random data to
ensure that each encrypted payload is different. The Rabbit case-study uses a commercial
implementation of SSL to illustrate the use of the protocol with HTTPS and web browsers.

Wrap-Up

SSL is everywhere, and its importance is greater now than ever. The reader is encouraged to study SSL for the design decisions made and the properties of the security it provides. The protocol is a solid example of good engineering, peer review, and how something can be widely known but still provides security after years of exposure. SSL will figure prominently throughout the rest of the text, as it is a primary example of the combination of security and flexibility that we will need to implement secure embedded applications.

In the next chapter, we will look at the role of security in embedded applications, and cover some of the issues involved in implementing security for a resource-constrained system. Following that, we will again take a look at SSL, and more specifically TLS, as it applies to wireless technologies.

CHAPTER 5

Embedded Security

In this chapter, we delve into the embedded realm. In this chapter, we will cover security from an embedded systems developer's point of view. First and foremost, the primary concept that will be introduced here and reinforced throughout the text is that the embedded systems we are covering are *NOT PCs or expensive devices running Windows or Linux* (see Figure 1). The personal computer and many expensive embedded devices have virtually infinite resources when compared to low-power, inexpensive, embedded platforms. The problem is that many security solutions rely on this almost limitless supply of memory and clock cycles. Cryptography is expensive, and the flexibility of some protocols (for example, allowing multiple algorithms to be used at any give time, such as SSL does) chews up a lot of code space and data memory. To further exacerbate the problem, embedded systems often require better response times than PC's, seeing as they are often in control of physical systems. Some security solutions that work well in resource-rich environments may simply not fit the criteria for many applications due to memory or clock cycle usage. For embedded machines, we need to design security based upon the target platform and application, as a general security solution will often be impractical or impossible to use.

As in the PC world, the security solution applied to any system depends on the application. For example, a web browser used for online shopping has much different requirements than a network-accessible payroll database (well, maybe they should, but that is for someone else to discuss). In the embedded world, the application matters even more, since there are very tangible tradeoffs in cost, performance, and security. We will introduce some examples here that will be used throughout the book to see the differences in requirements that different applications have.

In both PCs and embedded systems, the application matters, but for security on an embedded device, the target platform matters as well. A programmer in the PC world can treat all systems essentially the same (we handle discrepancies in platform performance with the "minimum requirements" disclaimer on the application's box). For embedded applications the developers cost, size, and performance of the platform all matter. In many cases, the budget of a project may limit the hardware that can be used. This does not mean we should abandon all hope of developing secure applications, but it does mean we need to be clever

Figure 1: PC vs. Microcontroller

in choosing what mechanisms and how we used them to secure our "lean-and-mean" applications. We will look at the tradeoffs of cost and performance when implementing different security options.

So at this point, you may be wondering: why not just go online and find a security package for your hardware and just buy and install it? The truth is, there are no general-purpose security packages available for a large number of embedded platforms. True, you can spend several hundred dollars on a device for embedded Intel Pentium 4's and just use off-the-shelf software solutions, but then again, you probably would not have picked up this book if you had that kind of money to spend. Furthermore, there are no embedded security standards in existence. This obviously leads to a dearth of off-the-shelf security packages for most embedded devices. For this reason, we will spend this chapter, as well as the majority of the rest of the text, looking at some of the protocols, algorithms, and mechanisms that are commonly used to secure digital transactions and see what we can do to squeeze them into the space we have.

Networked Embedded Systems and Resource Constraints

As the Internet grows, an increasing number of embedded systems are going online. From cell phones with web browsers and networked refrigerators to monitoring pipelines and factory floor remote control, the age of the network for embedded systems is upon us. It used to be that all the networking technology was reserved for expensive hardware and big corporations. The explosion of the "personal Internet" has brought the technology down in price to the point where embedded devices can utilize it effectively in new, novel applications that we could not even dream of fifteen years ago as seen in Figure 2. Wireless technologies, which we will cover in more depth in the next chapter, have created new challenges for security, since we are creating ever smaller and more mobile networked applications. As the concept of "Personal Area Networks" (PANS) becomes more and more of a reality, the need for security will be greater than ever, since everyone will essentially be broadcasting information to the world. New technologies like Bluetooth and ZigBee allow for the creation of endless numbers of networked devices. With all that information

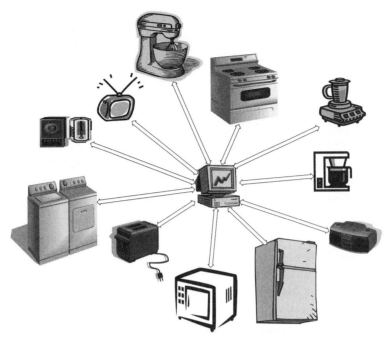

Figure 2: Networked Home Appliances

flying around, we are headed for interesting times, and security will need to figure prominently in this new reality.

The Internet has been restricted to PCs and large, relatively expensive hardware. As technology improves, we are starting to see some of the Internet technologies filter down to less expensive, smaller hardware. We can now realistically entertain the idea of massive swarms of tiny machines that self-organize into large ad hoc sensor networks, or complex control systems that require thousands of devices to intercommunicate. These applications require low-cost hardware to be economically feasible, and they need to be small. These types of applications will only work if the hardware being used is small enough for the application. A sensor network of millions of devices is not very useful (not to mention prohibitively expensive) if every device is the size of a baseball. For these types of applications to work, we need to look at limited-resource systems. This reveals yet another problem—how do we use the scant resources on these devices to implement our applications? These applications will almost definitely require some type of communications security, since they will need to communicate with one another—how do we fit the application and security into a device that likely has at most a couple hundred kilobytes of space for both data and code? The miniaturization of Internet hardware also has an interesting side effect in the embedded industry in general. Miniaturization costs a lot of money; as a result, many technologies are going to be kept closed and proprietary. Given that, how do we reconcile the utility of embedded applications with the need for security? We need to look at how security applies to these applications and how it can be implemented. Implementing security takes resources, and for these applications, cost is going to be an issue.

You may now be thinking, how does this apply to me? Well, think about it: If you could reduce the size of onboard memory on any embedded device, you could likely reduce cost since memory size and price are typically related (except in some cases usually with obscure, obsolete, or discontinued parts). If you could save just one dollar per device, it could save a lot of money if you were making hundreds or thousands of devices. Even with a small number of devices you might be able to reduce the size and cost of each one. Since professional developers continually have to balance costs with sound engineering (which are often at odds), security becomes a point of contention. Security is usually big and slow, and therefore expensive. The problem is that the lack of security can also be expensive, but in less clear ways. Many developers may be tempted to leave out security entirely in order to keep costs down and release products. It is even more likely that the engineer will not be part of the decision at all, a manager makes the decision and the engineer can only voice unheard complaints. But what if that engineer could show that the addition of security would not cost too much, since the security necessary is designed into the system and only uses a small portion of the available resources? This would almost completely eliminate the conundrum of whether an application should be secure—all applications can be secure. Engineers are happy with the sound engineering choice, and management is happy since they can pay a small amount to have the confidence that the application will be relatively secure.

But just how do we make a constrained system secure? Most security protocols and mechanisms are designed for systems with nearly limitless resources (at least as compared to our small embedded devices). They require fast processing and lots of memory (both for code and data). On top of that, security makes life difficult to design applications that need to react to real-world inputs. Many applications just do not have the time to wait for a generic security protocol to finish a communication before an action is required in response to some input. With the limited resources of our embedded systems, we have a serious barrier to implementing adequate security to our applications. How are we going to overcome this barrier and secure these types of applications?

One way to overcome the barrier of limited resources for security is to tailor the application to work with the security. There are obviously other ways to overcome the barrier, such as adding more resources, but we want to keep the price of the hardware down, so it is usually better to design the software to fit the desired hardware than the other way around. Our primary focus in this book is to provide a process of analyzing security protocols so that we may tailor our applications to the security desired, thereby freeing up resources and reducing the overall cost of the hardware required for those applications.

Embedded Security Design

The basic idea for tailoring applications to utilize security mechanisms and protocols effectively is to start off with a design that takes security into consideration *from the start of requirements gathering* as seen in Figure 3. Without starting the application with security in mind, the only option is to try and shoehorn in a security solution later on, which leads to additional headaches, schedule slips, and worst of all it can compromise the security of the application because the interaction between the bolt-on solution and the application may not be well defined. To begin the design of a secure embedded application, we need to look at all of the tradeoffs between performance, cost, and security. Unfortunately, these three concepts are almost always directly at odds with one another. More performance means the cost goes up, lowering the cost means lowering security and performance, and implementing higher security means performance will decrease.

With all these tradeoffs, where do we begin? Well, the best place to start is at the beginning—what is the application going to do, and how is it going to accomplish that goal? As an example, let's compare the security requirements for a couple of "big iron" (i.e., nonembedded with lots of resources) applications—a web store and a payroll system (see Figure 5). Obviously, these two applications have drastically different security needs (it should be obvious that both should be secure applications).

The web store expects interaction between its servers and hundreds or thousands of customers, essentially none of whom have any connection to the store, but whose identities are

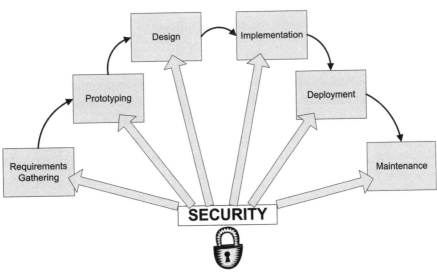

Figure 3: Steps to Designing in Security

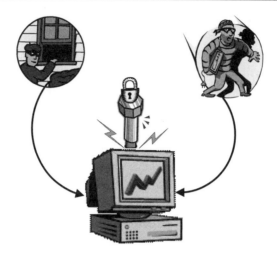

Sometimes it's easy to avoid bolt-on security

Figure 4: Bolt-on Security Isn't Always Ideal

	Webstore	Payroll System
Users	Thousands, anonymous	Dozens, with usernames
Security	Protect credit info	Protect all personal info
Data lifespan	Weeks	Years
Biggest threat	Hackers on Internet	Disgruntled Employee
Primary Interface	Web Browser	Proprietary
Performance	High	Low to Medium
Primary Security Protocols	HTTPS/SSL	SSH, AES

Figure 5: Web Store vs. Payroll System Security

important to keep safe. Obviously, the web store needs some way to protect online transactions, be able to authenticate itself to the customers, and be able to handle dealing with anonymous users through a web-based interface. The security here can be provided through a combination of public-key and symmetric encryption, along with some type of authentication system (such as the RSA authentication discussed in Chapter 1). As it turns out, this type of security is so common in networked applications that a generic security protocol was developed that provides almost all of the security that our example web-store application would need: SSL. The Secure Sockets Layer protocol (SSL, discussed at length previously in Chapter 4) is a generic security protocol that secures basic transactions over a network through a combination of public-key and symmetric encryption, message authentication through cryptographic hashing, and server authentication through a Public-Key Infrastructure (PKI). As we discussed, however, SSL is very generic, as it is designed to be blanket security for all network transactions, and the generic nature of the protocol means that it is big if fully implemented. We will discuss later how SSL can be used in parts to provide needed security in a small footprint, but for now we will look at our other example application.

The second example application is a payroll system for medium to large sized corporations. There are a number of people that require different levels of access to the system, such as human resources staff, the accounting department, as well as managers looking to hire new employees. Unlike the web store of our previous example, this application will likely be deployed on a corporate network that is shielded from the Internet, behind a firewall or other secure interface. The payroll system is also not subject to the anonymous user problem. Everyone accessing the system should be a known user, so that the accesses to the system can be logged, for legal or bookkeeping reasons.

The payroll system has a very different set of requirements for security from the online storefront. If we were to design these applications and try to apply a generic security mechanism to them after they were partially or fully implemented, we would run into a lot of trouble. The interaction between the security and the application does not fit very well, since the security is just "bolted on." No generic mechanism can even hope to capture the wide range of possible requirements for all applications. This is the primary reason why we should incorporate security into our designs at project inception, rather than try to add it in later.

The requirements analysis phase of the design of a networked application should include all of the security features that will be needed in the final product, and into the foreseeable future. This seems like a very tall order, but when you compare the cost of maintaining an insecure system (which will be vulnerable to many attacks, adding to the cost of maintenance) to the cost of a more complete design up-front, good design wins hands-down. Obviously, with any design, there are tradeoffs between cost, performance, time-to-market, and security—all of these aspects of the application work against one another. Higher security means higher cost and longer time-to-market. Higher performance often implies higher cost or lower security. Short time-to-market implies more cost, less performance, and less security.

In our example applications, we have some very definite tradeoffs. In the first example, the web store, time-to-market and performance are likely important, where cost and security are not as important (this is reality, security suffers because of economics)—see Figure 6. For this reason, the designer of our web store system will likely choose a generic security package to bolt into the application, which improves development time, but security may suffer from not taking the time to analyze the specific needs of the application (a web store will have the horsepower to use a generic solution). Depending on the security package,

1) Support thousands of anonymous users
2) Quick time-to-market
3) Medium to high security
4) High performance and throughput
5) Short-lived data support
6) Web-based interface

Figure 6: Web Store Requirements

1) Support dozens of users with audit trail
2) Value security over quick time-to-market
3) High level of security
4) Low to medium performance is acceptable
5) Keep and protect data for years
6) Console-based or proprietary interface

Figure 7: Payroll System Requirements

cost may or may not be affected. If the designer chooses an open-source package, the fee is likely minimal, possibly even zero. Companies that do this type of development are likely to experience problems down the road because they did not take the time to understand the special needs of their specific application. This is likely why we hear about companies "losing" thousands of credit card numbers or Social Security numbers to online bandits.

In our second application, the payroll system, we have a different set of requirements as seen in Figure 7. In this case, there is likely a system in place already, and the company commissioning the new system is looking to upgrade. In this case, cost and security are likely the most important considerations, with performance and time-to-market taking a back seat. That's not to say that the application should be slow and take forever to implement, but the application needs to be secure, and it does not generate revenue directly for its owners, so the cost and security are just more important. With this application, one or two dedicated engineers would likely concentrate on developing a thorough analysis of the security the system needs, taking into account access permission, the environment in which the application runs, the hardware on which the application will execute, and other factors, such as probable types of attacks and system failure. It is also a system that will handle a large amount of money, so there are likely practical and legal issues that need to be considered as well, such as how to store the information safely and restore it after a system failure, and keeping the system up-to-date with any laws and regulations in the region where the application is deployed.

Determining the security requirements for applications like those in our examples can be a daunting task. By definition, a successful attack is usually something you did not think

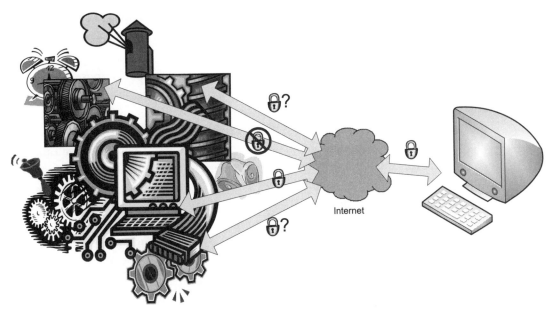

A complex system is much harder to secure than a simple one

Figure 8: Simple Makes Security Easier

about when you built the system (if you knowingly put a vulnerability into an application it is called a *backdoor*). A good security expert should be critical of every component of a system, since every piece, every user, every function can be a weak point that can be leveraged to exploit that system. So too must the system designer be watchful over the entire process, and not let ego or pressure sway his or her mind to let something "go." Sounds like a big problem, right? You probably have enough to worry about just getting the application to perform the way you want it to. What can we do to make our lives easier so that we can develop secure systems without sacrificing all of our free time? The answer is simple (sort of): Keep the application simple (see Figure 8). In the next section, we are going to discuss how keeping an application simple greatly reduces the work required to make that application secure.

The KISS Principle

Many readers have likely heard of the KISS principle—Keep It Simple, Stupid—referring to the fact that simple ideas are easier to grasp. Referring to electronics, it usually means that the simpler a system is, the easier it is to design, release, and maintain. Keeping it simple also has another benefit, which is why we have our own version of the KISS acronym: Keep It Simple and Secure. You see, a simple system is easier to understand. An easy-to-understand system is easy to make robust, since all the problems are likely known ahead of time or are easy to predict. Finally, as we mentioned before, robust systems are easier to secure, since there are fewer variables to keep track of when protecting resources.

One of the greatest pitfalls in securing a system is the "feature." For our purposes, we will throw out the marketing definition of and define a feature as any singular task that the application performs (i.e., adding an entry to a database or reading a value from a field in a form). Unfortunately, the most secure system would have no features at all. It would be trivial to secure, since there is nothing it can do anyway! However, if we understand that adding features increases the complexity of securing a system exponentially, we are going to be much more careful about the features that we choose to support. The reason for the exponential growth in features is due to the interactions between features, both intended, and more importantly, *unintended*. For example, with one feature, you only need to secure one thing—that feature. With two features, you have to secure *three* things: feature 1, feature 2, and the interaction between features 1 and 2. For three features, the count of things to secure increases to *seven* items: feature 1, feature 2, feature 3, features 1 and 2 together, features 1 and 3 together, features 2 and 3 together, and finally, the interaction of all three features together. For four and more features it gets much worse—see Figure 9.

It should be obvious that each feature needs to be secured, but why do we have to look at the interactions between features? Well, as an example, let's look at a possible attack against a system with two features, reading from a database entry and writing to a database entry. The database server is responsible for maintaining the integrity of entries before and after a write, and reads should not be able to affect the data in the entry. The write is secured against anyone stopping the write operation when it is only partially finished—the write operation has to be all-or-nothing. This makes sense—if the operation cannot be completed, then the application should not be writing what is potentially nonsensical data into the entry, thereby invalidating the integrity of the database. Now remember that a read operation cannot affect an entry value, it is strictly read-only. When run individually, there is no problem with either operation, reads have no affect on entry states, and writes will only complete if they can fully complete. Now let's put the operations together and see what happens. A write begins, and then an attacker initiates a read on the same entry. If the system designer did not pay attention to interactions between reads and writes, then it is conceivable that the read operation may have higher priority than a write. From a performance perspective, this makes sense: reads are a quick operation, and writes need extra verification to complete. Now if this is the case in our example, then the attacker can effectively prevent the database entry from being updated, which now violates the original stipulation that read operations cannot affect entries in the database—by preventing writes, the data becomes stale and is never updated, or updated too late as seen in Figure 10. In fact, it could be much worse if the interaction between the read and write operations was not clearly defined, since the read could clobber the write operation midexecution and

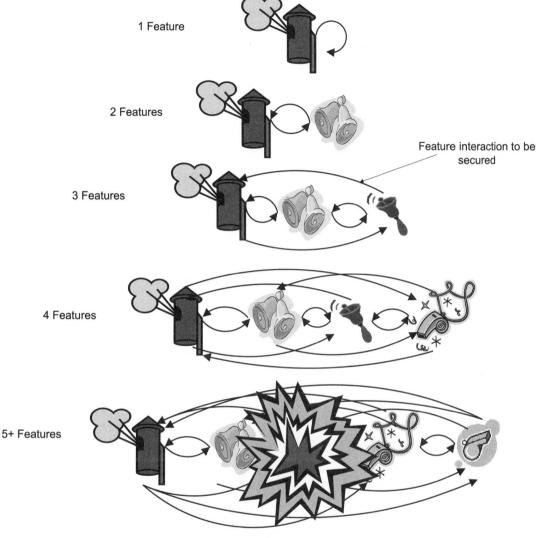

1 Feature

2 Features

Feature interaction to be secured

3 Features

4 Features

5+ Features

Figure 9: Feature Explosion

corrupt the data being written. These possibilities are both results of the unintended consequences of not paying attention to the interactions between features. In this simplistic example, the problems are easily seen—there are only 2 features, so there are only 3 combinations to worry about (read, write, read/write simultaneously). If the systems designer took into account the read/write interaction as well as securing the read and write operations, then the application would be much more robust and secure.

The problem comes in when we have so many features that the possible permutations of feature interactions grows so large it is not practical or feasible to model them all. When

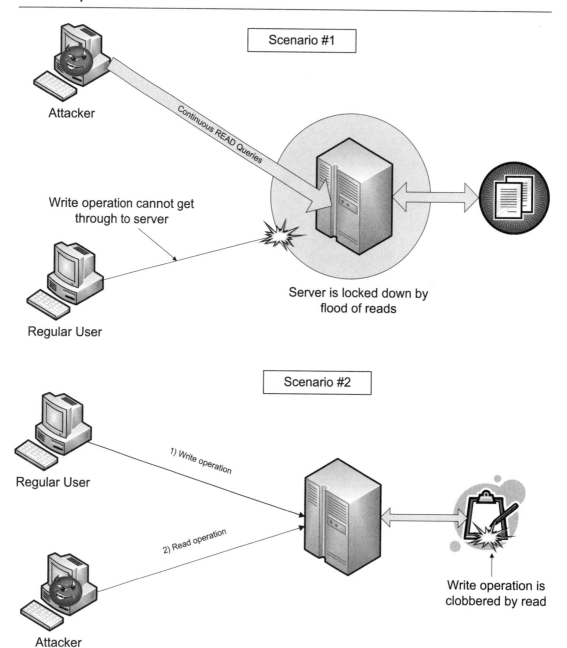

Figure 10: Inadvertent Read Behavior Prevents a Write from Happening

this happens, there is always the possibility of vulnerabilities existing deep within the application, often through some subtle combination of operations that the designers never considered. It is in our interest, therefore, that we keep the number of operations (and therefore features) to a minimum.

In the real world, this can be a very difficult task, as the sponsor of a particular project likely has a laundry list of features that "need" to be supported. In many cases, it may be possible to remove some features that really are not needed for the system to achieve its primary purpose. Careful analysis of the feature set keeping in mind the security ramifications of each additional feature can save time, money, and headaches in the future of the project. Any features that can be removed at design time should be, and if possible, *the system should be split into several different* components.

Modularity Is Key

Modularity is a cornerstone of good programming practice, and it can be extremely beneficial in securing an application. If you build small, robust, independent functional units, and test each one extensively, then you have some guarantee about the security of each of those units. Trying to guarantee anything about a monolithic application is a lesson in futility—see Figure 11.

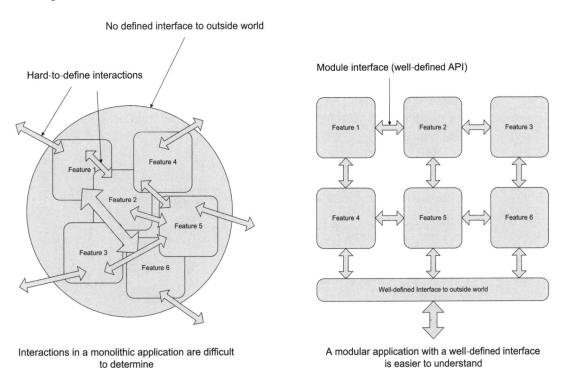

Interactions in a monolithic application are difficult to determine

A modular application with a well-defined interface is easier to understand

Figure 11: Monolithic vs. Modular Applications

A good modular design also has the benefit of flexibility. Implement as many independent units as you can, exporting a uniform interface, and you can mix and match the features depending on the application. For embedded systems with limited resources, this is a definite advantage, since devices can have targeted responsibilities without having too many extra features and their resource requirements. Once you have a number of robust, fully tested "feature units" and you are building your application, you can focus on securing the interactions between those units, with some guarantee that they will behave well.

To secure interactions between the modular units, it helps to have well-defined interfaces. There are several rules that you can use to keep your interfaces well-defined and less likely to be the cause of problems later. Some of these rules are listed in Figure 12.

The rules listed in Figure 12 are by no means a complete list of things you need to do to prevent errors from leading to a security breach, but they can help. For more detail, we expand on each of the rules:

1. If you are passing a buffer of information as a parameter to an interface function, always make sure that the buffer has an associated length, usually a separate parameter. The functional unit should also check that length and generate an error if the buffer length is not within a known range.

2. Avoid passing pointers (in C), if possible. Pointers by their very nature can be quite dangerous. If the pointer is corrupted (either by accident or by an attacker), it can lead to compromise of the system. If possible, limit the pointer to a particular region of memory, and abort an operation if that pointer ever refers to an address not in that region.

---

1. *Always* make sure buffers have an associated length when passing as a parameter. In the function, don't access the buffer past that length.

2. Avoid passing pointers (in C), if possible.

3. Avoid multiple levels of indirection.

4. Range-check *everything*.

5. Identify invariants and enforce them with assert.

Figure 12: API Rules

3. Avoid multiple levels of indirection. One level of indirection is hard to understand, two is even harder, and three is nearly impossible to comprehend in a complete sense. Sometimes two or more levels are required to simplify certain algorithms, but they should be carefully controlled using a memory protection scheme as mentioned in the previous item.

4. Range-check everything. This includes flags, numeric parameters, strings, and pointers. Any variable that can misbehave is a potential vulnerability, so define each variable's expected values at design time, and be sure to check that each variable never deviates from its expected set of values.

5. Assert is your friend. If you don't have the assert function, create your own that prints out information when something goes awry. Identify anything in your application that has a hard boundary, and protect that boundary with an assert. It may seem a bit cumbersome, but any assert failures you catch and fix during development and testing will prevent potential security breaches after deployment.

Now you may be asking, doesn't all this error checking make a program bigger? Well, this is indeed true, since range checking adds additional logic, limited use of pointers can cause extra code bloat, and passing a length along with a buffer is frequently cumbersome and uses up both code and data space. So how do we reconcile these needs with our constrained system?

The most obvious solution to the aforementioned issue is to remove all range checking, use pointers everywhere, and just assume that callers passing buffers all obey the rules. This is unfortunately the easiest solution to the problem, and therefore is likely one of the most commonly used, even though it can violate some basic rules of software engineering. An application that strips all of these features can still be fully functional, and is likely much smaller than its more reliable counterpart. Given this, we need to think about security very early on in the design phase of an application so that we can account for the additional resource requirements needed to implement that security. This is a good thought, but how do we really guarantee that a design will lead to a secure, reliable application that does not hog resources?

The best solution to making an application smaller is to remove features. Most marketing-types may balk at this, because the addition of extra features often sells more products, but there should be at least some level of compromise. Remember that each additional feature can add an exponential amount of complexity to the system, making it that much harder to have any guarantee that the system is secure. The most secure system has zero features,

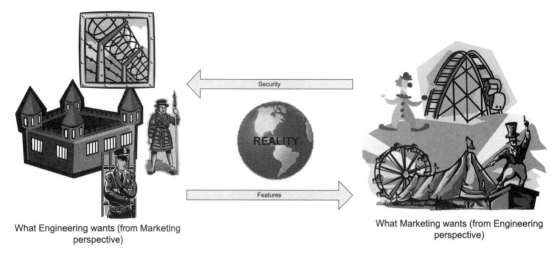

What Engineering wants (from Marketing perspective)

What Marketing wants (from Engineering perspective)

Figure 13: Marketing vs. Engineering

since there is nothing to exploit, but that system also has no practical purpose, since it cannot do anything useful either. In the middle between the extreme of zero-features and what marketing wants (everything) is the real world—see Figure 13. We should aim to develop applications with fewer features if they are to fit into a small, dedicated embedded system. Think about which features are absolutely necessary, and which are expendable. Try to define the features as well as possible, and nail down the requirements to the minimal set possible that still meets the goals of the project. Resist any attempts to add bells and whistles if you hope to have any level of security when the application is finished.

As an example, let's take a look at a sample application—a door monitoring system connected to an internal network. In our application, every door has a small monitoring device consisting of a microcontroller connected to a keypad and a magnetic sensor that determines whether the door is open or not. Each microcontroller is equipped with an Ethernet interface that is connected to an intranet within the building, and a centralized server provides data logging, control over the door locks, and other services (see Figure 14). In order to cut down on cost, the intranet is shared by the company's employees so that both they and the door monitoring system utilize the same network resources.

In our sample application, it is pretty clear what the embedded devices should be doing—checking whether the door is open or not, verifying key codes with the central server, and controlling the door locks. Any additional functionality should be considered potentially

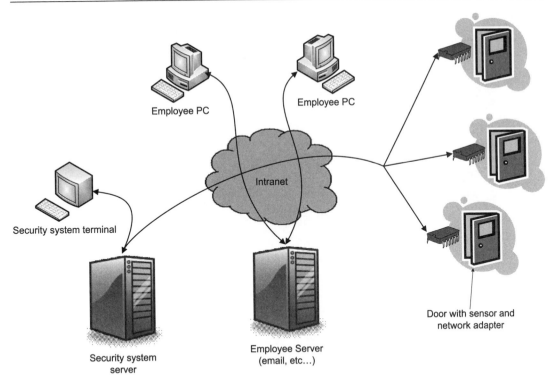

Figure 14: Door Monitoring System

detrimental to the security of the system, because those features would be strictly unnecessary for the proper functioning of the application. Sure, someone might want to add additional features such as displaying a message to the person opening a door or sending an email every time a certain employee enters the building, but features like these are not strictly necessary for the proper functioning of a door monitoring system. If they are not required by some regulation or for some safety reason, then they should probably be removed from the requirements of the application. Removing these extra features frees up additional resources for the application at each node, allowing for more boundary checking, free space to remove indirection, and has the benefit of reducing potential security holes.

Let's now take a look at a simple program that might run on one of the door monitoring nodes, in pseudo-C code:

```
main() {
    byte key_flag, alarm_flag, open_flag;
    byte key_value, key count;
    word open_count;
```

```
Key_flag = alarm_flag = open_flag = 0;

Key_value = key_count = 0;

Open_count = 0;


Init_Network();

Init_Door_Sensor();

while(1) {

  Key_value = get_key();

   if(key_value != -1) { // -1 indicates no key pressed.

      Key_array[key_count++] = key_value;

      if(key_count >= 5) {

         Have_key = 1;

   }

   }

   if(have_key) {

      Open_flag = check_key(key_array);

      Have_key = 0;

   }

   if(open_flag) {

      Open_Door();

      Open_flag = 0;

   }

   // Do other stuff (monitor door open/closed)

   }

 }
```

Listing 4.1—Sample Door-monitoring Main Program

In the sample program, the keypad is being continually monitored in the main loop. Once 5 numbers have been entered, it verifies the entered key by a call to *check_key*, a helper function that contacts the central server (over a secure channel—more on this in a moment) with the key, and awaits a response of true or false indicating whether the key code was valid or not. Several flags keep track of the state of the program, and determine whether to unlock the door or not.

Upon first glance, this program seems to be a perfectly reasonable solution to the problem, perhaps not the most elegant way it could be done, but functional. There do not seem to be any glaring reliability issues that would indicate any problems with security. However, we have some things to look at if we are doing an audit of this code to check for security and reliability.

First, we have several helper functions that are called from the main program. We need to know that these functions are well behaved, and that nothing we are doing is contributing to a potential security risk. The three functions in question are *get_key*, *check_key*, and *open_door*. Each function is connected to an external source that represents a potential security hole that we need to check. The get_key function retrieves the values of keys pressed on the numeric keypad by a user. The check_key function sends a completed key to the central server and awaits a response over the network. Finally, open_door is connected directly to the door lock, and will release the lock when called (assume that the lock mechanically returns to a locked state automatically once the door is opened and shuts again).

First, let's look at the get_key function. This is the one function with a direct connection to a user input device, and hence has the most potential for direct abuse. What would happen if the user were to mash a whole bunch of buttons at once? What if a key were held down or punched rapidly? Note that an attacker does not need to gain entrance to the room to accomplish his goal—maybe all he wants to do is prevent others from accessing the room for a period of time. If mashing all the buttons locks up the system and prevents other users from opening the door, for example, then it is relatively trivial for the attacker to achieve this goal. When developing this application, the get_key function may be accessible to us in code form or only as pre-compiled binary with an exposed API. In the latter case, all we can do is to check the interface documentation to be sure we are not violating any assumptions in calling the function and verify through some external means (such as black-box testing) that the function behaves the way we want it to. If we have access to the actual code of the function, then we can check to be sure the code does not do anything that would lend itself to malfunctioning if some unexpected input were encountered.

Assume that we have written our own get_key function that utilizes an internal buffer along with an interrupt service routine that updates a locally-accessible buffer whenever a key is pressed on the keypad:

```
// ISR - key_value contains the value of the key following the interrupt firing

void key_isr() {

    Key_buf[cur_key_pos++] = key_value;

}
```

```
// get_key function - returns a word containing the value of the next key press
//following the previous call to get_key

word get_key() {

    word ret_val = -1; // Default = no key available

    If(cur_key_pos > sizeof(key_buf)) {

        // circular buffer - reset current position in buffer

        Cur_key_pos = 0;

    }

    if(last_key_pos != cur_key_pos) {

        ++last_key_pos;

        If(last_key_pos > sizeof(key_buf)) {

            // circular buffer

            Last_key_pos = 0;

        }

        ret_val = key_buf[last_key_pos];

    }

    return ret_val;

}
```

Listing 4.2—Sample Code for Keypad Interface

Our hypothetical application has the keypad wired to an external interrupt that fires whenever a key is pressed. This seems reasonable enough, except that there is a potential issue if a key is held or pressed repeatedly. If an interrupt is allowed to preempt a previous interrupt, then this is a poor implementation, since the behavior is undefined if keys are pressed simultaneously. For this reason, let's assume that either interrupts cannot be interrupted, or a semaphore or other type of lock is used on the global Key_buf buffer. Even if this is true, however, there is still the issue of the ISR firing every time a key is pressed—this could potentially lock up the device if a key is held down. More importantly, look at the implementation closely—the reset of the cur_key_pos counter is in the user-called function. If the ISR is firing rapidly, it is quite possible that the get_key function will not execute before the counter is incremented past the end of the buffer.

If this code were buried deep within a helper function in the application, it may never be caught, since there does not appear to be any problem with it. There really is no logical problem with the code, which makes it even harder to spot the issue. The majority of the time, the circular buffer and its counter will be well behaved. In this case, however, holding a key down will likely cause an overrun of the Key_buf buffer, after which the code starts stomping on memory. This in and of itself is undesirable behavior, but take a look back at the main implementation in Listing 4.1. The variable open_flag exists somewhere in the data space of the device, either on a stack or in a static data area somewhere. If our misbehaving ISR was to continue writing non-zero values to memory, there is a good possibility that the open_flag variable would be overwritten at some point. Once execution returns to the main program loop, the flag would be non-zero since there is no place it is set to 0 except after opening the door. If you haven't figured it out by now, the door is now open, completely bypassing the actual security key check. All someone has to do to open the door is to hold any key (other than zero) for a while (probably not long), and the door will open.

This example is a little scary since the bug was hard to spot and the impact of the security breach is significant (unauthorized access to the room). The buffer seemed to be well behaved, since in normal thinking, the ISR fires once, and the get_key function is called from the main loop soon afterward, resetting the counter. It isn't until we start thinking of the things we aren't thinking about (also called "outside the box" thinking) that we can see that there is such a serious issue. We will look at how we can fix this to avoid this type of problem later on, but first we will look at the code a little more.

Now that we are suspicious of this implementation, are there any other problems that could lead to potential security risks? Well, there is the issue of sending the key over the network to the central server over the intranet, but we have already indicated that this will be done in a secure manner using SSL or some other cryptography. However, an attacker still may be able to get at the value of the key without having to decipher the encrypted message. If we look at the both the main program and the get_key function, we see that there are 2 different buffers that store the key information while a key is being entered and again when the key is being processed. Neither of these buffers is cleared explicitly, and they only are overwritten when a new key is entered. This means that the key information is stored in plain (unencrypted) form in 2 different places in the memory of the node device. The key may not be sent explicitly over the network in plaintext, but the memory of the device may be accessible.

In some networked embedded systems, a resident operating system or monitoring program provides services that allow remote debugging or firmware updates. It is not inconceivable that the debugging services may not have been turned off before deployment of the system, which would allow anyone with network access a possible back-door into the system. Depending on the services available, this oversight might be a serious breach in security. For example, if random memory can be changed the attacker could just change the flag in memory to open the door. More likely, the services are more restricted, but it may be possible to get a memory dump. One method may be to attack the target and to get it to crash and send a core dump to the server. Another possibility is that one of the debugging services left open may allow for reading memory directly, in which case the attacker could do some reconnaissance. The attacker could punch buttons on the keypad of the device that is vulnerable and then return to his computer and do a memory dump to see where the pattern he entered was located in memory. Then all he would need to do is to wait for a legitimate user to open the door to do another memory dump and presto, he has a valid key. Granted, this situation would require the attacker to have access to the same network that the door nodes are on, but this may not be too unrealistic—imagine a disgruntled employee that wanted access to a server room or HR records for nefarious purposes. In fact, the general belief (with ample evidence) is that the majority of security issues for a company come from within the organization.

The case of the "inside man" brings up another issue—the communications used by the embedded devices to contact the server and receive authorization. Let's specifically take a look at the receiving end of the application: the nodes. In the following listing, we have a static buffer and a global flag that is used to control the door mechanism (open_flag). The buffer is used to receive messages from the server via a secure SSL connection. The messages, for efficiency, are always the same length (100 bytes), so the buffer is that length as well, since we want to conserve our static data space.

```
// Static data
char buffer[100];
char open_flag = 0;


// Read from a particular SSL socket into a static buffer
void read_message(SSL_sock_t ssl_socket) {
    // Reads message from the SSL socket into the static buffer if
able
    if(ssl_ready(ssl_socket)) {
      // Writes into buffer current SSL record contents
      ssl_read(ssl_socket, buffer);
    }
  … // Do other stuff
}


// Reads from an SSL socket into a buffer (passed via pointer)
void ssl_read(SSL_sock_t ssl_socket, char * buf) {
    int bytes_ready;
```

```
    bytes_ready = ssl_bytesready(ssl_socket);

    ssl_sock_read(ssl_socket, buf, bytes_ready);

}
```

Listing 4.3—Static Data and Receive Functions

In the listing, you can plainly see that the buffer and open_flag are adjacent in the program. Some compilers may try to pack these variables into different places in memory, but in all likelihood, they will be allocated sequentially. At this point, you should already be a bit suspicious of the close proximity of the buffer to the all-important open_flag. Suspend that thought for a moment, and let's take a look at the actual code. We have two functions, which is probably overkill for this application, but maybe there is a use for the ssl_read helper function other than where it is used in read_message. In any case, take a look at what happens when an SSL record has been received from the server (assume the SSL connection is already established) and read_message is called. First, read_message checks to be sure the SSL socket is open and ready with information before proceeding. Since we have a record, we then call ssl_read with our SSL socket and our static buffer as parameters.

In ssl_read, the logic retrieves the amount of available data from ssl_bytesready, which in this case returns the number of bytes that are available to read from the internal SSL buffer. A quick look at the API documentation (pretend here) and we see that the function will only return up to the total number of bytes in the *current* SSL record. So far, so good, we now know that ssl_bytesready will return no more bytes than are available in the current record. We know that all the messages from the server will be 100 bytes (as was specified in our design), so everything should be just fine. But hold on, what if the SSL record is larger than 100 bytes? Well, this would probably be an error on the server end, since SSL does a pretty good job of protecting the client. An attacker would be hard-pressed to send anything that the SSL implementation would accept once the session has started, so any packets not from the server would be rejected by the SSL implementation. There is still a bug, but it does not allow the attacker access to the secured room. It just so happens that the nature of the SSL protocol prevents the overflow bug in ssl_read from being a major issue (unless the server starts sending messages larger than 100 bytes!), but what if for code space reasons, something less than SSL was used? If the application used a simple shared-key symmetric cryptography to ensure messages, it might be possible for the attacker to append information to the fixed-length message (updating the TCP/IP packet, for example) which would not be noticed until after the decryption was done. Assuming that the decryption operation is in the read_message function, it becomes almost trivial to utilize the buffer overflow bug to set the open flag and open the door. This is yet another example of a subtle bug that may never be caught in routine testing, but could lead to a big problem later, albeit in this case, it would require a relatively major change to the application.

The preceding example is somewhat contrived and greatly simplified from what would likely be encountered in a real application, but it illustrates the point that no amount of planning can account for all the potential security holes present in an application. So how do we deal with bugs that are essentially impossible to predict? In the next section we will start to look at some engineering techniques that help to alleviate the impact of unknown issues by reducing the number of unknowns through simplification and by boxing-in the areas we have less understanding of.

Well-Defined Interfaces

One of the key features of a robust modular system is the existence of well-defined and restricted interfaces between modular units. A well-defined interface includes well-understood pre and post conditions, definite limits on variable-length parameters, no "extra" parameters, and it has good documentation so that it can be understood by the developer working with it. The last point here is often overlooked, since there is often an implicit assumption that anyone using the API can figure it out from a terse function reference. There is also an implicit assumption that the API functions are well behaved and generally can be trusted. Do not trust API functions. Well, trust them, but only as much as you have to. It never hurts to wrap a call to a function that is not yours with some extra checking to be safe. There are a few things that you can look for to be sure that an API function is well-defined and will likely behave itself. These features are also important when designing your own API functions and modular applications.

First and foremost, check pointers. A well-defined API function should utilize type checking (if available) to protect the pointer in the function, such as utilizing the *const* pointer modifier available in C++. This is where C can get you into big trouble, since type checking is a little thin (OK, practically nonexistent), and many C implementations do not support *const*. If the function takes a void* as a pointer, there better be a good reason for it or it's a good reason to be suspicious.

If possible, utilize functions that require a length parameter if a buffer is to be passed in as a parameter. The presence of the length parameter indicates that the function is not relying on some inherent property of the buffer to know when to stop writing to it. For example, the C standard library comes with a number of functions for string and buffer manipulation that rely on the null-terminated property of C strings. There is obviously a problem if a buffer happens to not be null-terminated. The function will simply keep going until it encounters a null somewhere else, which may be thousands of bytes away. In a PC environment, this buffer overflow condition can pop an attacker into an administrator or super-user mode of the resident operating system, if it happens in just the right place. In an embedded system with a limited RTOS or no operating system at all, the administrator/super-user problem is less of an issue, but as was shown in the example in the previous section, it can let the attacker get what he wants. Here is a prime example of where good engineering practices can save you—by using functions that don't allow buffer overflow, you have not only made the system more reliable, but also more secure.

Buffer and pointer handling are not the only areas where good software engineering helps. Look for your functions to have a single, well-defined purpose. If you find yourself writing a kitchen-sink, monolithic, thousand-line function, stop. You may even want to throw it out and start from scratch (yeah—try convincing your boss of that one—but seriously, it may be worth it). Such a large function is not only hard to debug and difficult to understand, but there are very likely some states in that function that do things you don't want an attacker (or a plain user) to do. The problem is that the security holes (and bugs) are always in the code you understand the *least* (see Figure 15). Aim to break your code into manageable pieces that you can understand, and limit their interfaces so you don't get bogged down in function interdependencies. The more you understand the code you write and all that it can do (intended or not), the more secure your application will inherently be.

At this point, you may be asking yourself why such modularity is being recommended for resource-constrained systems. Don't all those extra function calls bog down performance and use up a lot of code space? While it is true that the overhead of the additional calls will slow down some of the performance of certain parts of the application and use up code

Figure 15: What You Cannot See Can Hurt You

space, remember that we are looking for modular, understandable code. If you are intelligent about how those functions are implemented, and keep critical sections to a minimum of modules, you can write a program that is just as small and quick as its monolithic counterpart, and it is easier to understand and more secure to boot! When we look at some sample applications later, we will discuss some of the ways we can shrink our code whilst retaining the modularity that is key to our application's reliability.

So, the point of all this software engineering discussion is that an unintelligible program cannot possibly be secure. Even if it was secure by some crazy improbable coincidence, there would be no way of knowing. The reality is that there is no way of knowing if an application is secure or not, there are only varying degrees of confidence (see Figure 16). While it is extremely easy to prove a system insecure (just attack it in some known way), it is impossible to do the opposite—how do you test the infinite number of possible inputs to a program (don't forget all the unintended inputs!)? Moving forward, we are going to start looking at ways to pick apart, deconstruct, and shrink protocols so that we can have the benefits of secure computing on our tiny embedded machines. As we look at these methods of shoehorning protocols and algorithms into machines that are too small, remember that you should attempt to understand as much about what you are doing as possible, because it is inevitable that the one function you don't understand will be the source of all your problems.

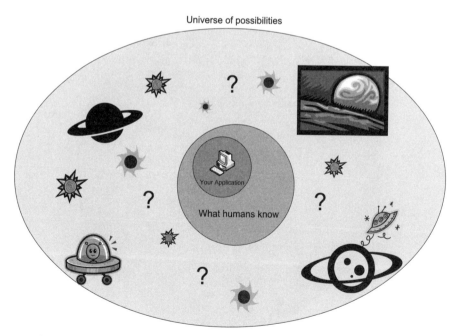

Figure 16: There Is a Universe of Possibilities Beyond What We Know

Pick and Pull

This section of the book is probably more appropriately titled "things protocol designers never intended" or "other nasty things you can do to your program that would void your warrantee if one existed." The goal here is not to look at any specific protocol in detail, but rather to give you the crowbar (and the courage) to pry into those nasty protocols and take them down to their most basic pieces to rebuild them anew with the focused, singular purpose of serving your application. We have covered the nitty-gritty details of some real protocols already, and we will cover more in chapters to come, but here we want to look at the properties of protocols that will allow us to get what we want out of them.

RULE #1: Removing a feature can be safe, but make sure you understand it first.

One of the first things you realize when you dig into most protocol implementations is that you are almost always dealing with a state machine of some type. It may be explicit in the code, or implied by the code logic, but it is there. After all, a protocol is nothing more than a series of steps to achieve some purpose, and some of those steps can be repeated (in other words, a classic state machine). If you can start to understand how the protocol works and what the states are in the state machine (it helps to look at the actual protocol specification), then you can start to see what parts are not strictly necessary. One easy way to spot "features" that you can remove is to look through the specification for the words "optional" and "should." Many Internet protocols are described in Request for Comments documents (RFCs) and those words have special meanings. The presence of those specific words is usually an indication that the protocol does not strictly need whatever is being described and that feature can probably be ditched. In an embedded application, we don't want a Mercedes-Benz, we want a dune buggy. Scrap everything that isn't strictly needed for operation or safety (see Figure 17).

Horsepower is good for some
applications

Resource-constrained applications
are not flashy

Figure 17: Resource-Constrained Applications Are Utilitarian

If you are really clever, and you feel like a rebel, you can also look at some of those things in the specification that are labeled with "must" and see if it really "must" have that feature. Often, the "required" features are needed to meet a protocol specification, but are not strictly necessary for the protocol to operate. As an example, to be TLS-compliant (Transport Layer Security, which is essentially the public standard for SSL) you "must" support the 3DES encryption algorithm. This does not mean that a TLS implementation will not function without 3DES, it just implies that every "official" implementation has a guarantee of being able to speak that algorithm. The reality is that you could ditch 3DES for AES, which is smaller and faster, and you would probably have as good or better security. This example is pretty clear-cut in that it is fairly obvious that you could build a TLS implementation without 3DES that would still function with a large number of other TLS and SSL implementations, but it wouldn't "be" TLS since it doesn't meet the specification. Other cases may not be so clear-cut, and if in doubt, don't cut it out—it isn't worth it to risk complete lack of security to save a few hundred bytes of code space. Try to look for other places to trim the fat.

RULE #2: Don't ever try to outsmart cryptographers. They are smarter than you.

As a general rule, it is usually a good idea to always assume cryptographers are smarter than you (unless you are a professional cryptographer). These people have PhD's in math or computer science (often both) and get paid a lot of money to develop ways to hide information so that other people can't read it—it's their *job*. Fortunately, they are usually nice enough to let us in on their discoveries so that we can put all that brainpower to use by implementing those methods and publishing them. Sometimes there is a license or royalty

Don't reinvent the wheel...

There are plenty of wheels out there to choose from!

Figure 18: Don't Reinvent the Wheel

attached, but for the most part, you can get your hands on at least a pseudo-code implementation of any number of powerful encryption algorithms. *Don't reinvent the wheel—just use one someone else created.*

There is no shame in using existing algorithms and protocols to secure your application, but *do not*, under any circumstances, try to optimize or remove parts of an encryption scheme (unless you are, in fact, a cryptographer, then by all means go ahead). The algorithms that you hear about, 3DES, AES, RSA, these are all proven, tested algorithms that rely on certain characteristics of the math on which they are based. If you think you can, go into the algorithm and remove a couple of rounds—say, you have just undermined the security of that algorithm. One of the ways cryptanalysts (cryptographers who try to break cryptographic algorithms) attack an algorithm is by first reducing the number of iterations an algorithm is run over some data to see if a pattern emerges. If you go in and reduce that number to gain a few clocks, you have just made the attacker's job a whole lot easier.

The rule applies also when attempting to roll your own cryptography. The algorithms that are most commonly used in practice are popular because they have been excruciatingly designed to be secure. Some algorithms have taken years of design and analysis to reach the point where they are trusted enough to be universally adopted. All these algorithms are publicly available and have been subject to extensive scrutiny. Some mathematicians make a career out of attacking algorithms and publishing their results. Even worse, there are plenty of people with less honorable intentions that want to break these algorithms for nefarious purposes. Unless you are a genius or are extremely lucky, the chances are that any algorithm you produce will have a fatal flaw that will allow someone to attack your implementation with ease. And don't think that hiding your algorithm will work either. "Security by obscurity" is a tempting method to keep your data protected, but as has been proven with many systems throughout history, it is not too difficult to discover how a cipher works without knowing its actual implementation. It may seem like a good idea to implement your own algorithm, since you will understand how it works and can optimize it to your liking, and there are no sticky licensing or rights issues, but it is really better to find an existing algorithm. Plus, existing algorithms are compatible with other systems, and it makes communications easier to implement, since you can use off-the-shelf applications and mechanisms with your application.

RULE #3: *You do not need everything that every protocol offers.*

In the world of the PC, you have nearly infinite resources with which to play. Since the PC is the current dominant entry point to the Internet, it makes sense that the myriad of protocols in existence today are designed around the PC's basic specifications: loads of memory, a giant hard disk, and clock speeds measured in gigahertz. Well, if you are in the market of building and deploying thousands or millions of units, you do not want to pay for all that extra horsepower. Unfortunately, protocol designers working on the PC platform are inherently sloppy when it comes to resource usage. They tend to add in a number of bells and whistles that make the protocol more *generally* useful, but do not really add to the core functionality provided by the protocol. What the embedded designer should be looking for are the parts of protocols that are *specifically* useful for a particular application. In the SSL/TLS example mentioned earlier, the removal of 3DES from your implementation is one such feature that can be omitted from an implementation to conserve resources. A PC implementation of TLS will include dozens of different cryptographic algorithms, providing a whole host of possibilities. A PC is a general-purpose machine so it needs to be ready for anything. Generally speaking, an embedded application is dedicated to a specific purpose and does not need all the extra functionality—it only has to be compatible with the other parts of the application.

RULE #4: Make sure you only apply "safe" optimizations to security protocol implementations.

When you are implementing a security protocol for a resource-constrained system, it will often be tempting to try and optimize parts of the protocol that appear to be slowing things down or using up too much memory. Be extremely careful in what you optimize and how you do it. When implementing an algorithm like AES, you may notice that the internal loop that conducts the "rounds" of the algorithm, iterating over the text being encrypted a number of times. It would be very tempting to try and speed up the algorithm by reducing the loop count, sure it wouldn't work with other AES implementations, but that's OK, right? All you have to do is copy the optimized algorithm to both ends of your application (i.e. client and server) and it will work just fine since the algorithm is the same on both sides. Well, you need to refer back to rule #2 (above), since you are trying to outsmart the people who designed AES in the first place. Every tiny detail of the algorithm has been designed to provide the security guaranteed by using that particular algorithm. All the constant values seen in any cryptographic algorithm implementation are carefully selected for certain mathematical properties and should not be touched.

What can we do, then, if optimizations are so dangerous? What we need is a concept of "safe" optimizations, transformations that can be applied to the code without breaking any of the assumptions inherent in the algorithm. One such example is the use of bit-masks in place of the modulus operator. If you notice that there is a place in code that is always using modular arithmetic, it may be possible to use bit-masks to solve the problem. Typically on any machine (especially small embedded controllers) a bit-mask operation is significantly faster than the equivalent modulus operation. On a processor that does not have native instructions for division, there is no contest. The only downside to the bit-mask trick is that the modulus must be a power of two. Fortunately, some algorithms already use powers of two since they are easy for binary computers to handle. For example, the RC4 encryption algorithm uses what is essentially a hash table of 256 bytes. In the "hash" function of RC4, there is a modulus with 256 so that the index value ends up pointing to only one of the spots in the table. If you know your binary math, you will see that "index MOD 256" is equivalent to "index AND FF," where AND is a binary bitwise "and" operation and FF is 255 in hexadecimal. In RC4, this operation is applied to every single byte that is encrypted, so you can see that reducing a division operation to a simple AND will have significant impact on the performance of the algorithm. Also note that the RC4 specification did not change, since we did not change what was being done, just how it was being done. As an example of what the actual code for this optimization would look like, refer to Chapter 8, where we will look at this optimization in detail.

RULE #5: Simple is better.

As has been mentioned before, simplicity makes an application easier to understand, and therefore easier to make secure. Don't overcomplicate your application with lots of extra protocols and features. List the requirements of the application and cut out any that are not *absolutely necessary*. Being an embedded systems designer, you should already be doing this, so this is a very complementary operation. Keep the feature set down, and you will have a smaller application. Having a smaller application means you can study it more and make it more robust and secure. Avoid feature creep at all costs as well. If new features are being added or old features are being expanded as development continues, you are introducing a delta to your existing code base that makes it even more difficult to analyze what you have done. On top of that, the extra time taken up by implementing the new features could have been used for tightening up what you had already implemented.

Justification

So why do we have to do so much analysis and picking apart of protocols? Well, in a word, cost. If you have plenty of money to throw at the problem, then just throw more hardware at the problem. There are plenty of off-the-shelf products that will allow you to build a secure embedded system around some expensive embedded hardware that has hundreds of megabytes of storage and RAM and a processor running in gigahertz. If you have that kind of money for your application, then you probably wouldn't have picked up this book in the first place. If you have a tight budget, or have to build thousands or millions of devices, then every byte you can shave off of the hardware requirements, every additional megahertz you don't need is a big deal. We want to be able to build devices that have all the properties of their more expensive brethren for a fraction of the cost to open up new opportunities and new markets. In order to achieve this, we have to be a little creative about how we go about implementing our applications. We can't be wasteful like PC developers (have you seen the requirements for running an O/S like Windows these days? You may remember that a megabyte or two of RAM was plenty enough for an old DOS machine . . .). Instead, we have to conserve every shred of resources. This chapter has introduced some of the concepts that we can use to cut back on the waste of modern protocols and algorithms, but the rest of the book will cover more specific examples of application design and features that we can exploit to keep resource usage (and cost) down, while retaining at least a reasonable amount of security. There are no "embedded security" protocols or methods right now. Everyone just uses what is out there for the PC/Internet world and tries to push it all into the embedded space. Hopefully the information in this chapter and those following will assist you in being a little more proactive about designing a good, secure embedded system without resorting to buying more and more hardware.

Wrap-Up

This chapter has focused on the ideas behind the "miniaturization" of security and the challenges faced by embedded systems engineers when implementing security. Throughout this and previous chapters, the idea of wireless technologies has been a consistent thread. As we have mentioned, embedded wireless technologies make embedded security more important than ever. In the next chapter, we will look at wireless communications in the context of embedded systems. Wireless technologies embody all of the concepts discussed thus far, and we will see how they all come together in wireless protocols. Wireless technologies for embedded systems are still extremely new (for example, the ZigBee standard isn't really even a standard yet—it is still being worked on), so we will cover the protocols as they exist now and attempt to generalize the concepts of security and wireless technologies together so the ideas will apply to emerging technologies.

CHAPTER 6

Wireless

Wireless Technologies

Wireless applications essentially combine all of the reasons we need security for embedded systems with limited resources. By definition, many wireless devices will require limited resources, because they will be designed to run on batteries or in environments where available resources must be given to the communications hardware. In the world of inexpensive embedded systems (we exclude inexpensive consumer products from this category; we are referring more to industrial-type controllers), wireless technologies are only just starting to make inroads into the industry. Sure, cellular phones and PDA's have had wireless technologies built-in for some time, but those implementations are specialized and not widely available. For this reason, and the fact that many embedded wireless communications implementations are still jealously guarded by the companies that produce them, there are not a large number of available real-world embedded wireless devices to discuss. However, as wireless technologies decrease in cost and their implementations increase in number, we will begin to see more and more wireless devices available to smaller companies and hobbyists. Wireless is a radical change in the way devices communicate, so this infusion of technology will present some interesting challenges to embedded developers. In this chapter, we will look at a few of the most popular and up-and-coming wireless technologies and the security implications of changing our preferred communications medium from wires to the air.

As we discussed in Chapter 3, wired network hardware has dominated the Internet landscape for years. Wireless technologies predate the internet (think radio), but until recently, the technology had lagged behind wired technologies in providing the connectivity required by modern networked and Internet applications. Wi-Fi,[1] which follows the IEEE 802.11 standard protocols, has been around for a while, but the technology was typically reserved for consumer applications and PC's—the infrastructure and physical characteristics of the hardware just did not work with many embedded applications. We are just now seeing the 802.11 protocols make their way into the lowest reaches of the embedded realm, but Wi-Fi

---

[1] The term "Wi-Fi" is a trademark of the Wi-Fi Alliance, www.wifialliance.com.

is by no means the only wireless technology available. In this chapter, we are going to discuss Wi-Fi, cellular technologies, ZigBee,[2] Bluetooth,[3] and other protocols that can be used for embedded applications. Before we get into the details, though, let's take a brief look at some of the other technologies for wireless communications.

While the Wi-Fi protocols dominate the computer industry, cellular communications is probably the most recognizable form of wireless technology. Cellular communications technologies have been around for a long time, and in some sense, cellular was one of the original "embedded" wireless protocols, since cell phones pretty much embody the principles of an embedded paradigm. Cellular communications is relatively old in the world of technology, but it was optimized for a single application—voice-based telecommunications. Recently, technologies such as GPRS have expanded cellular communications to more general data-driven applications, but it has been a little slow in coming, and any applications usually need to have the backing of one of the large cellular telecommunications companies. That being said, the cellular networking infrastructure is stable and ubiquitous and can be put to great use in applications where none of the other wireless protocols would be effective.

Cellular technology allows for devices to be connected to the global Internet from just about anywhere, but sometimes the scale (and cost) of cellular isn't needed or even justifiable for an embedded application. In recent years, two technologies have risen to the forefront of wireless communications specifically for the embedded world. The first of these to make a name for itself was Bluetooth, which is the technology behind hands-free mobile phone headsets, PDA keyboards, and a host of other consumer devices. However, perhaps even more exciting than Bluetooth for the embedded systems industry is the rise of ZigBee. ZigBee is an 802 protocol, like Ethernet and Wi-Fi (802.15.4 to be exact). It was specifically designed for low-power embedded applications. Bluetooth has some of the same properties, but has found a different niche as a sort of wireless USB protocol, where ZigBee is more reminiscent of a wireless version of good old RS232 serial port. ZigBee is a relatively new protocol (the standard is still in the process of being finalized), but it is already being put to use in some exciting applications. In this chapter, we will look at both Bluetooth and ZigBee, and discuss the security features built right into both of these useful technologies.

Cellular Technologies

Cellular wireless technologies were created for mobile telephone communications but, like their wired counterparts, have diversified and evolved into general-purpose communications

---

[2] "ZigBee" is a trademark of the ZigBee Alliance, www.zigbee.org.
[3] "Bluetooth" is a trademark of the Bluetooth Special Interest Group, www.bluetooth.com.

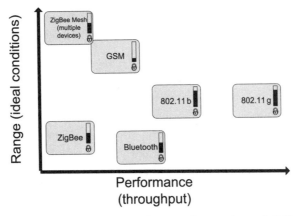

Figure 1: Comparison of Wi-Fi, ZigBee, Bluetooth, and Celluar/GSM

technologies. A few technologies are of interest when discussing the connection between cellular networks and digital communications, including GSM[4] (Global System for Mobile communications, the base technology for a majority of cellular communications) and GPRS[5] (General Packet Radio Service), which adds data transfer capabilities to GSM and allows for services like text messaging and data communications. There are numerous other cellular wireless technologies, but we will keep our discussion to GPRS/GSM because of its widespread use.

One of the largest barriers to using cellular technologies for inexpensive wireless communications (in our case, for embedded control applications) is that cellular networks are difficult to get on to, usually requiring a partnership with the organization that owns the network. For a lot of applications, the cost of this may not be practical. However, there do exist companies that do that part for you, and you can buy GPRS/GSM modems that will allow your application to be connected to a cellular network (the modem vendor will usually have a partnership with at least one or two carriers).

The closed nature of cellular networks makes security a difficult problem. The GSM and GPRS technologies have security built into their specifications, but the methods used are not the best. Poor encryption algorithms and questionable security design considerations mean that cellular communications may not be as secure as they could be. If you are going to use

---

[4] Originally, GSM referred to "Groupe Speciale Mobile", a European project for developing mobile technologies. Now it is managed and trademarked by The GSM Association (GSMA), www.gsmworld.com.

[5] GPRS and GSM have been expanded into new technologies that provide higher data rates and other services, first EDGE and later 3GSM. We focus on GPRS, since the new technologies just expand its basic functionality.

GPRS/GSM as a communications medium, it is recommended that you use a higher-level security protocol (SSL is a good choice) on top of the communications channel.

Cellular networking allows for a couple of features that are interesting for embedded applications. The networks are available nearly everywhere, so a cellular-enabled device would have a network connection nearly anywhere, and cellular networks are very good at providing roaming connections, so devices can move around. However, for a large number of embedded control applications, cellular technology is probably overkill. If the embedded device is in a warehouse somewhere and does not move around too much, but needs wireless connectivity, since wires are difficult to run, cellular is probably too slow (dial-up modem speeds are normal) or expensive. For this reason, we leave our discussion of cellular technologies and look at some more practical wireless technologies for limited-resource applications (and as a bonus, they all happen to be generally easier to secure than GPRS/GSM).

802.11 (Wi-Fi)

The wireless communications technologies usually referred to by the (trademarked) term "Wi-Fi" are those technologies based off of the IEEE 802.11 standards. Intended as a general wireless communications protocol (think of Ethernet without wires), 802.11 implementations are by far the most common form of wireless communication between PCs. Wi-Fi is characterized by having a medium range of communications capability (as compared to cellular) with a very large (relative) data rate.

802.11 wireless is a heavy-duty wireless protocol, supporting speeds that rival wired Ethernet (802.11b is capable of 11Mbits/second, and 802.11g is capable of 54Mbps). Its designers recognized the need for security early on and included a security protocol in the original specification: Wired Equivalent Privacy, or WEP. However, WEP was inherently flawed, due to the use of stream ciphers without accounting for some of the important issues inherent in using stream ciphers (see our discussion on RC4 and stream cipher caveats in Chapter 3). One of the major issues was the use of short initialization vectors and infrequent changes of the master RC4 keys. Today, there are numerous implementations that show WEP can be broken in (literally) seconds on modern hardware. Despite this obvious flaw (which is basically a total lack of security), WEP will still stop eavesdroppers that don't know the technical details (i.e. stupid criminals), but it will not stop anyone else, so it should really never be used. One problem, however, was that thousands of (or more) applications were developed for Wi-Fi systems that used and relied on WEP, often in hardware or difficult-to-update firmware. As a result, fixing WEP was not an easy proposition, so a compromise was developed, Wi-Fi Protected Access (WPA).

WPA improved upon WEP by addressing the most grievous flaws exhibited by the original protocol, but retained a level of backward-compatibility that allowed WPA to easily be implemented for most systems that previously relied on WEP. One of the major improve-ments was to up the effective key size of 40 bits provided by WEP to a full 128 bits. Another major improvement was the use of automated dynamic key management to assure that the stream cipher keys (for RC4 in WEP) were changed on a regular basis to avoid the problem of key reuse (in Chapter 3 we noted that if the same stream cipher key is used for two different messages, the plaintext was easily recoverable from the ciphertext without actually knowing the key). Authentication was also a problem with WEP, which used the WEP key itself, so WPA uses a separate authentication protocol (there are actually several protocols that can be used, as we will discuss).

WPA, though far superior to WEP (and still widely used), carries some of the concerns about WEP, since it provides backward-compatibility with WEP hardware. For this reason, follow on improvement to WPA was introduced by the Wi-Fi Alliance, creatively named WPA2,[6] which is based on the IEEE 802.11i standard. Both WPA and WPA2 improve security by moving to AES (and also supporting various key sizes larger than 128 bits as required by the US government), and differentiating between personal and enterprise networks, which have different requirements. WPA-Personal is designed for home net-works, where authentication is not as important as for a business. WPA-Enterprise is basically the same for protecting data (uses AES), but it provides much stronger authenti-cation using a centralized, managed authentication server. The additional management requirements for the centralized server make it too cumbersome for home use, hence the split.

WPA Key Management

WPA and WPA2 (both Personal and Enterprise) utilize a key management mechanism called the Temporal Key Integrity Protocol, or TKIP. TKIP provides the dynamic key management that addressed the key reuse problems in WEP. TKIP is primarily used for WPA-Personal now, since it is based on the RC4 cipher, rather than the (assumed) more secure AES. In order to make deployment easier, WPA-Personal supports what is called a Pre-Shared Key, or PSK. The terms WPA-TKIP or WPA-PSK are often used to refer to WPA-Personal or WPA-Personal. For WPA2-Enterprise, the preferred method is called the CBC-MAC[7] Protocol (CCMP), based upon AES. In order to be compliant, WPA implemen-tations must support TKIP, and WPA2 implementations must support both TKIP (for Personal) and CCMP (for Enterprise).

---

[6] WPA and WPA2 are trademarks of the Wi-Fi Alliance.
[7] Counter-Mode Cipher Block Chaining (CBC) with Message Authentication Code (MAC). CBC is a method of using the AES block cipher as a stream cipher.

WPA Authentication

Authentication in WPA (and WPA2) is utilized to prevent unauthorized connections to a network and to help mitigate threats from rogue access points (so-called "evil twins" that trick the client into believing it has connected to the correct network). For WPA-Personal, authentication is not strictly required because of the work required to manage an authentication server. For WPA-Personal, the PSK is usually considered enough for authenticating home wireless networks, but the stronger methods could be used if desired.

For Enterprise authentication, both WPA and WPA2 utilize the same basic framework: 802.1X/EAP. The 802.1X protocol (note that the "X" is really an X, not a placeholder for a number) is part of the IEEE standards for managing both wired and wireless networks, and defines a secure transport mechanism for EAP messages. EAP, the Extensible Authentication Protocol, provides the basis for authentication in both WPA and WPA2.

EAP was originally implemented for WPA2, and at the time the only EAP-variant required for Wi-Fi Alliance compliance (needed to use the trademarked Wi-Fi name and logo on a product) was called EAP-TLS, TLS being the Transport Layer Security protocol we discussed in Chapter 4. However, the number of EAP variants has grown considerably, as various vendors have created their own mechanisms, all slightly different. Now, there are a total of 5 variants required to be compliant: EAP-TLS, EAP-TTLS/MSCHAPv2[8] (TTLS is simply "Tunnelled TLS"), PEAPv0/MSCHAPv2 (Protected EAP, which establishes a TLS connection over which EAP methods are used), PEAPv1/EAP-GTC (an EAP variant developed Cisco), and EAP-SIM which is essentially authentication using SIM cards for the telecom industry. In any case, Wi-Fi authentication is a dynamic and complex field and keeping up with it can be quite a challenge (by the time you are reading this it is likely that there have been a number of new protocols added and compliance requirements have changed). If you are interested in learning more about the wide array of authentication mechanisms for Wi-Fi, there are numerous online resources and there are even a few books on the subject. A good place to start is the Wi-Fi Alliance website itself: www.wi-fi.org. Figure 2 shows the relationships between the different authentication mechanisms used by Wi-Fi.

Drowning in Acronyms

The alphabet soup of Wi-Fi authentication protocols begs one important question: What do we need to support? Well, like we have talked about before, we need to adapt the protocol to our application. If you are implementing the latest and greatest consumer gadget and

---

[8] MSCHAP is a variant of the PPP CHAP protocol developed by Microsoft. MSCHAPv2 is described in RFC 2759.

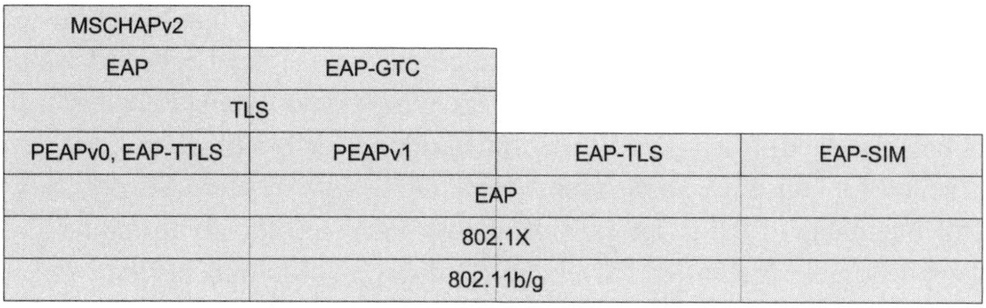

MSCHAPv2			
EAP	EAP-GTC		
TLS			
PEAPv0, EAP-TTLS	PEAPv1	EAP-TLS	EAP-SIM
EAP			
802.1X			
802.11b/g			

Figure 2: Wi-Fi Authentication Mechanisms

you must be compliant with every wireless access point under the sun, you will likely need to implement most or all of the protocols and mechanisms (or more) described above (you probably want the Wi-Fi logo too, and you definitely need more information than this book provides). If you are working on a proprietary solution for a specific purpose and you can control what access points are used and just want to have some level of security for your embedded devices, then you can probably scale back to a lower level of authentication. In fact, for many embedded applications (especially those with strict budget limits), WPA-PSK may be sufficient. The full WPA-Enterprise authentication suite was designed for large organizations with numerous high-power devices such as laptops and expensive PDA's that need to be continuously updated. The level of security provided by an authentication server is probably overkill for an application that monitors the output of an oil well (for example).

Another important point that has not been addressed is the fact that Wi-Fi is a high-throughput, and therefore high-power wireless protocol. It is very likely that if you are developing an application using a low-power inexpensive microcontroller (which is why you picked up this book, right?), the amount of power required for the 802.11 radio probably exceeds your requirements. In other words, if you are looking to add wireless connectivity to an inexpensive embedded device, you probably do not want Wi-Fi.

Do You Really Need 802.11?

The extensive requirements of Wi-Fi, both in software support and in power consumption, make Wi-Fi a less attractive option for limited-resource systems. We mention it because it is so prevalent and it forms a majority of digital wireless connectivity today. It is possible to implement 802.11 wireless for inexpensive systems, but the functionality will likely need to be reduced to meet the system specs. Fortunately, more than a few people recognized the need for wireless protocols that provide connectivity without the resource requirements of full Wi-Fi. Two protocols have risen in recent years that promise the level of connectivity needed by low-power and inexpensive devices without having to support numerous security protocols and without having the power consumption associated with the higher bandwidth

802.11-based protocols. The first of these is Bluetooth, a standard that has come to be a household word due to its widespread use in mobile telephone headsets and various other consumer devices. The second protocol is a relative newcomer (the standard is still in the process of being ratified) is ZigBee. Where Bluetooth provides a medium level of throughput and is suited for consumer applications (think of it as a wireless USB port as we mentioned at the beginning of the chapter), ZigBee is tailored specifically for embedded industrial applications which often have radically different requirements than consumer applications. We will finish out the chapter by looking at both Bluetooth and ZigBee and how their inherent security properties can be put to use in embedded applications.

Bluetooth

Named after a relatively obscure Scandinavian king, Bluetooth was one of the first wireless protocols to address the power consumption issues that are inherent in battery-powered consumer devices. By reducing the bandwidth and range requirements, the Bluetooth protocol lends itself to battery-powered applications that require a moderate level of throughput, such as wireless headsets for mobile phones and input devices (such as keyboards and mice) for PDAs. Originally developed by a consortium of technology corporations, Bluetooth has been widely adapted by vendors of consumer gadgets. Driven by widespread use, the Bluetooth physical layer specification was adapted by the IEEE to develop the 802.15.1 standard.

The security of Bluetooth, as with all the wireless protocols we are discussing in this chapter, is designed right into the standard itself. The standard was developed and is controlled by the Bluetooth Special Interest Group (www.bluetooth.com), and the security is based on a 3-mode model, with an unsecured mode, a "service level" secured mode, and a link-level secured mode (the entire connection is secured). According to the Bluetooth SIG, all known attacks against the Bluetooth protocol are actually against specific implementations and the protocol itself is secure.

The security of Bluetooth uses the concept of two separate keys, an authentication key and an encryption key. The authentication key is the master key, and encryption keys are regenerated with each new session. A random number, generated for each transaction, adds additional security. The basic cipher for data protection and the authentication mechanism are described in detail in the Bluetooth specification, should you choose to implement it yourself. However, there are a large number of vendors that supply complete Bluetooth solutions on a single chip, some of which may implement the security in hardware (or the entire Bluetooth stack, as National Semiconductor does with their Simply Blue modules). Due to the availability of such hardware (the Simply Blue modules sell for less than $30 each at the time of this writing), there is very little need to understand the Bluetooth stack in any detail, unless you want to try and implement your own Bluetooth solution.

It should suffice for our discussion to say that there are no known serious attacks on the protocol itself, but various implementations may be vulnerable to a few attacks, referred to as "bluejacking," "bluebugging," and "bluesnarfing."[9] All of these attacks relate to the ability of an attacker to connect to a Bluetooth device (in most instances a mobile phone with Bluetooth) without the knowledge of the device user. The bluejacking attack simply involves the sending of an unwanted message to the device user, which could be used to trick the user into providing sensitive information to the attacker (phishing). The other attacks involve the ability of an attacker to access the contents of a device, either being able to execute commands (bluebugging) or to download data from the device (bluesnarfing). In any case, these attacks require the attacker to be in close proximity (within a few meters) unless they have the equipment to boost the Bluetooth protocol's range. Apparently these issues have been addressed in newer implementations of the protocol but as always, it is a good idea to keep up with current developments, as a devastating attack can always be right around the corner.

Bluetooth provides a decent midrange protocol for embedded systems that need a moderate level of throughput, but it is a complex protocol (the specification is well over 1200 pages long), and the cost of a dedicated controller unit may be prohibitive depending on the application. In the next section, we will look at a relative newcomer to the wireless arena, the ZigBee protocol. Designed around the IEEE 802.15.4 low-power radio standard, ZigBee aims to be the go-to standard for industrial wireless communication where throughput is less of an issue, and flexibility, power consumption, and cost are primary concerns.

ZigBee

At the low end of the power-requirement spectrum for wireless devices, ZigBee[10] is the equivalent of a dripping faucet when compared to the garden hose of Bluetooth or the fire hose of 802.11 (in the case of 802.11g, a water cannon used for putting out aircraft fires), as seen in Figure 3. ZigBee is a relatively new standard developed and maintained by the ZigBee Alliance. Like the Wi-Fi Alliance and the Bluetooth SIG, the ZigBee Alliance is a consortium of corporations that all utilize the protocol. ZigBee, as of this writing, has not been widely deployed due to it being so new. However, all indications point to ZigBee making a large splash in the industrial controls arena, since it is specifically tailored to such applications. As we mentioned before, ZigBee is characterized by low power consumption (able to run on batteries for extended periods of time due to the low duty cycle of its radio), low system resource requirements, and low throughput. The bandwidth of ZigBee is

---

[9] Terms and definitions for Bluetooth attacks adapted from the Bluetooth SIG overview of Bluetooth security (http://www.bluetooth.com/Bluetooth/Learn/Security/)

[10] Parts of the description of ZigBee in the section are adapted from the ZigBee Alliance ZigBee FAQ at www.zigbee.org.

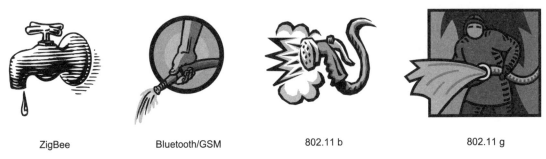

| ZigBee | Bluetooth/GSM | 802.11 b | 802.11 g |

Figure 3: Throughput Comparison

comparable to that of a dial-up modem (up to 250KB/s), but for the applications it was designed for, this will not be an issue. At first glance, ZigBee may seem very similar to Bluetooth, and in some respects it is, but the target applications for each protocol have led to some significant design differences. Bluetooth, being designed for consumer applications, focuses on higher bandwidth and convenience. ZigBee, on the other hand, is primarily concerned with flexibility of the network (ZigBee supports several network topologies that increase reliability of the entire network—we will look at these in a minute) and conservation of resources, especially power consumption. Geared toward industrial automation (as opposed to consumer connectivity like Bluetooth), ZigBee is definitely an industrial standard.

Like the other wireless protocols we have been discussing this chapter, ZigBee has security built into its specification. Some of the more interesting features of the security inherent in ZigBee have to do with self-healing mesh networks. ZigBee allows for thousands of nodes to be included in a single network, usually referred to as a Personal Area Network (PAN), which was coined to describe the networks specified in the IEEE 802.15 standards (ZigBee radios conform to 802.15.4). ZigBee actually supports several different network topologies, including mesh and clusters, as seen in Figure 4. The interesting thing about some of the topologies supported by ZigBee is that they are self-healing, in that the network is resilient and can deal with nodes coming in and out of the network, as would be expected in a noisy (RF noise) industrial environment. This self-healing property makes redundancy very easy to implement, and this can directly translate into a more secure application. Picture an attacker taking out nodes in a sensor network to prevent certain information to be collected. In a Wi-Fi network, once a node is dropped, it must reestablish the connection, including any authentication required. The attacker could take out a few key nodes and the entire network goes down. With a ZigBee network, there can be many more nodes (ZigBee is cheaper to implement) and if any nodes are dropped due to tampering, the network continues to function. The basic functionality of ZigBee is setup to use the radio as little as possible to conserve power, so the end result is that the nodes are usually "down" anyway.

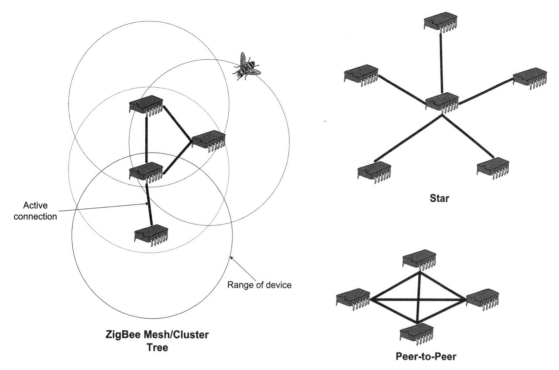

Figure 4: ZigBee Topologies

Without the need to keep close synchronization (as is the case with Wi-Fi, for example), ZigBee networks can deal with attacks inherently. Next we are going to look at some of the special needs of particular topologies and the security protocol built into the standard.

The network topologies supported by ZigBee vary, but depending on the type of network being deployed there are varying security considerations. In some ZigBee topologies, all nodes are considered equal and the network is basically an ad hoc peer-to-peer configuration. In some of the other topologies (star and tree forms), there must be a *coordinator* that facilitates the network. In such a topology, the end-nodes can be of reduced functionality to save cost, but there is an inherent problem: If the coordinator is disabled, reduced-functionality end nodes cannot reestablish the network. One way to get around this is to provide redundant full-function nodes that can all serve as a coordinator (this is a good idea for reliability, not just security). ZigBee nodes can also function as routers, directing communications between nodes that may not be able to communicate directly (as would be the case if the distance between those nodes was too great, but the router node bisected the path between them). If a ZigBee node is acting as a router, it is important that any information being passed along cannot be compromised if the router node is compromised.

The ZigBee protocol provides low-level security for communications between individual nodes using AES and a message authentication code scheme, but this only protects the data between nodes, not on the nodes themselves. Normally this would be OK if the nodes in question were functioning as end-nodes, but if they are functioning as coordinators or routers, it may be desirable to use a higher-level security scheme to protect that data. This presents another interesting challenge due to the limited bandwidth of ZigBee and the extremely limited resources that are likely to be found on ZigBee devices. A full-blown security protocol like SSL will probably not work because of the extra overhead (especially when you consider running RSA on a ZigBee node). In fact, a number of protocols, like IPSEC, do not even make sense, since ZigBee networks are not based upon TCP/IP. What would make more sense would actually be to use a simple AES-based scheme similar to what is done later in the PIC case study. Authentication could be provided through a password or key that is passed in encrypted form to the end device.

ZigBee is an exciting new technology that promises to bring a level of connectivity never before seen. Depending on how it is deployed, it may also represent a security challenge unlike any we have ever seen. What will be important to remember as ZigBee devices are deployed in factories and homes around the world is that these devices will probably not support a level of security equal to that of full-blown networking. Due to the extremely limited resources, some devices may have no security at all, so it will be vital to keep a constant vigil on what information is being sent over these networks. For these reasons, ZigBee will likely have a place in developing sensor networks and monitoring technologies, but it remains to be seen whether it will be used for anything more.

Wireless Technologies and the Future

The future of embedded technology lies side by side with the future of wireless. As wireless technologies are improved and are reduced in cost to the point they are as cheap to add to a system as a PIC, we will start to see an explosion of new applications as literally everything is connected to the global Internet. Embedded wireless communications will eventually make science fiction into reality as "smart" paint for bridges and buildings give real-time data on structural conditions and swarms of wirelessly connected micro-sensors show us the internal structure of tornados and hurricanes. Along with this diffusion of technology into every niche of our existences, security takes on a whole new meaning.

It is likely that engineers and corporations will push wireless technology to its boundaries, and security will be of the utmost importance. The lesson to be learned here is that you should consider security when choosing a technology for your particular application. When it comes to security, it never hurts to err on the side of caution and use far more computing power than you need to be sure that you can support the security you need both at deployment and years down the road.

Wrap-Up

This chapter has been a quick tour of a few of the most relevant wireless communications technologies for embedded applications, but is by no means a complete treatment of the subject. There are dozens of other standard technologies available and likely thousands of proprietary and custom solutions. In the next chapter, we are going to look at some higher-level protocols that will have some use to embedded applications designers, and we will continue to look at the role of security and how it applies to embedded applications.

CHAPTER 7

Application-Layer and Client/Server Protocols

Introduction

An increasing number of embedded applications use the Internet as a medium for communications, synchronization, or the applications provide services for the Internet and its users. The Internet has traditionally been dominated by the classic client/server model of communications, and it remains the standard model today, though some inroads are being made by new technologies, notably peer-to-peer networking. In this chapter we will discuss the most prevalent client/server protocols on the Internet, open them up, and see what we can do to make them secure, but still small enough for modest hardware. Most of these protocols require more horsepower than a standard low-end PIC, but surprisingly, they can often be implemented in a small enough footprint (with security, of course!) to work on a very modest 8-bit platform with memory measured in *kilo*bytes!

Client/server applications are characterized by a server program that is continuously available and waiting for incoming communications on a network, and a client program that runs on remote hardware (also connected to the network), that can connect to the server at any time to request a service or resource, such as a web page to display to a user or a remote process to handle a problem that is too complex for the client hardware.

In the embedded world, applications can be set up so that the embedded hardware can be a server or a client, and can be configured to communicate with other embedded devices, PC's, and anything else that can be connected to the network. Constrained-resource systems designers can utilize the flexibility of client/server protocols and mechanisms to implement intuitive web-based interfaces for everyday items, and numerous other possibilities. When these types of applications become commonplace (in some places they already are!), security becomes a concern for everyone. In the world of the Internet-enabled toaster, do you really want some 12-year-old hacker to be able to control that toaster from 1000 miles away?

In order to study the security properties of these higher-level protocols and mechanisms, we will pull apart several examples as we did with the lower-level protocols to see how the internals work, what inherent security (and lack of security) is present, and what we can cut out to effectively shrink our application without compromising the integrity or security of the entire system. We will start our tour of client/server applications and protocols with the biggest and most recognizable of them all—the World Wide Web.

The World Wide Web

Unless you have been living in a hole on some obscure undiscovered island, you know about the Web. Most people have heard of the Web, even in the most unexpected places. The universal adoption of the Web has led to a revolution in communications. Things that used to take hours of research to discover can now be learned with a single Google search. The Web and online services like eBay are turning mom-and-pop businesses into million-dollar companies. Without the Web, the Internet would likely not have as much impact as it does. The Web is easy to use, even for nontechnical people, allowing those from all walks of life to access a world of information.

This unprecedented level of global communication and information sharing comes at a cost—the world is smaller, and there is money to be made, so the ugly side of humanity rears its ugly head, and Internet crime has become a multibillion-dollar problem. Take Nigeria, for example. A decent percentage of their economy consists of sending illicit Spam emails to the United States and other wealthy countries. Russian hackers make millions breaking into Internet systems and wreaking havoc. Domestic hackers work to control thousands of machines (often referred to as zombies) to launch Denial of Service attacks in order to extort money out of corporations that rely on their Web presence. These crimes are made possible largely through the existence of the Web.

So what can we do? We can let the politicians sort out the social and economic problems, but we can also protect our Internet applications from many of the security pitfalls that come from being online. It simply requires that we outthink and out-plan those wishing to attack us. If we do our due diligence up front, we will not have to worry as much once the application is deployed.

The Web is obviously not the only client/server application that we can look at, there are many others that are used extensively on the Internet and can be of use for many embedded applications. Some of the protocols we will cover include the Hypertext Transfer Protocol (HTTP, the basis for the Web), the File Transfer Protocol (FTP), the Simple Mail Transport Protocol (SMTP, or email), the Simple Network Management Protocol (SNMP), as well as a few others. The Web is the most approachable from a security standpoint since most people are at least peripherally aware that some type of security exists when they are making online purchases. For this reason, we will cover the Web and its basic protocol (HTTP) first.

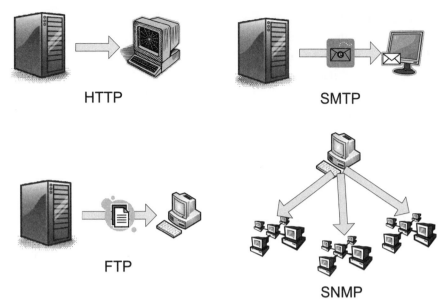

Figure 1: Common Application-Layer Protocols

Web-Based Interfaces

The Web is by far the most used Internet "application," and that makes it the largest target for malicious hackers. For this reason, the security on the Web is some of the most mature of any network application. Various types of authentication, encryption, and integrity checking are used in combination to make the Web secure. Due to the intuitive interfaces that can be provided by Web technologies (HTML, Javascript, etc.), it makes sense to use those technologies with embedded systems, especially for those systems that require interaction with nontechnically-inclined humans. Many embedded devices now support web servers and clients, and the improvements in low-cost hardware have allowed some of that technology to filter down into nontraditional applications. One of the problems with this, however, is that security is often sacrificed to allow for these applications to exist. Indeed, some developers believe that security for web interfaces is basically impossible on 8-bit microprocessors. However, that is not the case, as there are low-cost components that come with fairly complete security options.

In this section, we will delve into some of the base technologies behind web-based interfaces and see what types of security are built-in, what can be added on, and most importantly, what can be left out without compromising the security of the application.

To start, we will discuss the basic protocol that the Web is built upon, the Hypertext Transfer Protocol (HTTP). As you may know, HTTP is the high-level transport protocol for web pages written in HTML (the Hypertext Mark-up Language). HTTP controls the handling of requesting, serving, and processing of HTML files through a client/server model that includes methods for requesting resources, providing resources, and handling error conditions in a controlled manner. HTTP is a text-based protocol that allows for web browsers and servers to send requests for resources and responses to requests over a standard TCP/IP socket connection, or any other network communications protocol that allows for remote network communications using text.

At its deepest levels, HTTP is essentially a very simple protocol. The protocol basically consists of a request sent by the client for a particular resource. The server, upon parsing the request, will look for the corresponding resource and provide a response to the client with the resource (typically a text file consisting of HTML-formatted text and images) if that resource exists. If the resource does not exist, or some other problem occurred during the transmission or parsing and handling of the request, HTTP defines a host of error conditions and corresponding codes to indicate to the client that the resource is unavailable or somehow unreachable.

One feature of HTTP that is of particular interest in our discussions about security is HTTP authentication. HTTP authentication comes in two different flavors: plain, which just asks (via a plaintext request to the client) for a username and password, and hashed authentication, which uses a cryptographic hash algorithm to provide slightly more security than the plain authentication. The hash authentication mechanism is of some interest to the embedded systems designer because it represents probably the simplest secure method for authentication (plain authentication is not really secure at all since the password is sent over the network in plaintext—it can be of use with a protocol like SSL, however). HTTP hash authentication is useful, since it only requires a small hash implementation (traditionally either MD5 or SHA-1) to provide authentication. The authentication method utilizes a mechanism called challenge-response, which is common in typical client-server applications. The basis of the mechanism is that the password being used for authentication is known by both the client (entered by the user) and the server (stored as part of the application's static data) and to authenticate a message does not require the password to actually travel over the connection.

In practice, the way this mechanism works is through a random number that is passed from the server to the client called the *challenge*. The client, rather than passing the password back to the server, takes the challenge number, hashes it with the password provided by the user on the client end, and passes only the hash back to the server. This hash is called the *response*. The server knows both the challenge number and the password, so it can calculate the same hash as the client and compare it to the one sent over the network. If the hashes match, then there is reasonable certainty that the client user provided the appropriate password. As long as the server sends a different challenge for each client connection (not just every client), then there is no chance of a replay attack, where a malicious user sends the same hash to the server at a later time to attempt an unauthorized connection (the challenge sent to the malicious user will be different than the one used to compute the original hash by the client). The mechanism is possibly susceptible to a man-in-the-middle attack—if the attacker is able to spoof the address of the client and server machines, intercept the response from the client by pretending to be the server, and then pass the client hash to the server by pretending to be the client, the attacker would appear to be the client to the server (see Figure 2). However, if the application has good control over the addressing on the network, making it relatively easy to prevent spoofing, then the authentication can be fairly robust (assuming the hash algorithm being used is still secure, which may not be the case for MD5 and SHA-1 anymore).

Assuming the target system can support an HTTP server, it is not unreasonable to expect that the system could also support a hash algorithm like SHA-1. Hash algorithms, as was mentioned before, are easily optimized for a resource-constrained environment. Utilizing the hash authentication mechanism can provide reasonable security without the overhead of a complete security solution such as SSL.

It may at first appear that the plaintext authentication mechanism is completely useless, since the password is sent plaintext over the network, allowing anyone with network access to view the unencrypted message. However, it can be combined with other

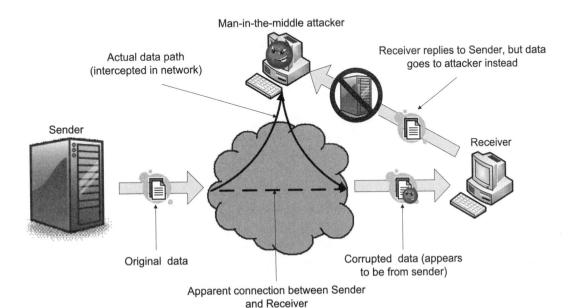

Figure 2: Man-in-the-Middle Attack

mechanisms to provide adequate security. For example, when used with a lower-level security protocol, such as SSL, it provides a very efficient and effective mode of authentication, since the password being sent is encrypted by the underlying mechanism. We originally looked at plaintext HTTP authentication in Chapter 4 when we looked more closely at the SSL protocol. SSL is a fantastic foundation for building secure web interfaces for embedded devices, provided the hardware you use can support the additional overhead.

Now that we have identified HTTP authentication as a possible security mechanism for a secure embedded application, what can we do to tailor it for the application to give us more resources for the application itself? We can look first at the hash method to use, to see if one algorithm is better suited to a limited-resource platform. For example, MD5 has a smaller default hash size than SHA-1, 16 bytes for MD5 versus 20 bytes for SHA-1. By choosing MD5 over SHA-1, we can save a few bytes of buffer space. This may not be important for some applications, but if we are shoehorning an HTTP server into a total of 64KB of RAM on a true 8-bit micro, a few extra bytes could mean a lot.

We will continue our look at web interfaces by looking first at web server applications, where a client user can connect to the embedded device using a basic web browser application. Following the web server discussion, we will look at making the embedded application support a web client, allowing the device to initiate communications with some centralized networked web server. We will also look at how we can choose to support either a web server or a web client, and the situations in which both would be needed. In all cases, security is a concern, and we will look how each implementation of server, client, and the combination of the two affects the built-in HTTP security mechanisms, and what we can do to be sure that the application is robust and secure in all cases.

Server-Side HTTP Web Interfaces

For many applications, it will be desirable to implement a web server on the target device, so a user on a PC somewhere on the network can use a natural client application—a simple web browser. Almost all web browsers today support various types of security mechanisms, from the simplest plaintext HTTP authentication to the full secure channels provided by HTTPS (HTTP secured using SSL). For now, we are going to focus on plain HTTP and the built-in security provided by the protocol itself, without delving into the specifics of coordinating HTTP and a lower-level mechanism such as SSL. For HTTP, the basic built-in security is provided by the various forms of HTTP authentication, as discussed above. We can use either plaintext authentication or hash authentication, with the latter being far more secure. If we are going to implement a web server for our embedded application, then we need to look at how to efficiently generate a challenge message for the connecting clients so that the authentication can be reliable and secure.

In order to assure that the challenge to a connecting client is secure, we need to have some level of confidence that the challenge itself (a random number) has a very low probability of being predicted by an attacker. If the generated challenge is not a cryptographically secure number, an attacker can more easily produce a hash to provide to the server application that will fool the server into accepting an invalid user, or worse, attain the actual hidden password. If the attacker can predict challenges that will be sent in the future, it would be possible to do a dictionary attack on an older message (with a known challenge), derive the password, and then produce a "correct" hash when a predicted challenge is sent. For this reason, we need to be sure that the number we generate is as unpredictable as possible. We can utilize a combination of network traffic and hardware properties such as interrupts firing, serial port traffic, and incoming network packet sizes to feed into a pseudo-random number generator to provide fairly secure challenges. However, none of these sources alone is very good to use, since each source on its own can be manipulated to be more predictable. The worst possible random number seeds are fixed numbers and the system clock; the latter was actually part of what caused SSL version 2 to be so easily broken.

It is a good idea to spend a lot of time making sure that you have a suitable PRNG (pseudo-random number generator) and decent seed material, preferably from a physical source (electrical noise can be used as a seed, for example). Having a good random number for the challenge is a good start to securing an HTTP server. As we will see with the HTTP client in the next section, a PC or server machine will be generating the challenge, so there is much higher probability of the number being cryptographically secure (a more powerful system has more physical inputs to utilize in generating unpredictable randomness). This could be an argument against implementing a server on the embedded device, but as we have mentioned before, it really depends on the application. If the application does not need the utmost in security (for example, if the application resides on a physically secure and isolated network) then the embedded device should be adequate enough to generate sufficient challenges.

There are not a great number of things that we can do to make the server implementation of the challenge less resource-hungry. Unfortunately, there are not a whole lot of features that can be removed from the server side and still have it be compatible with a wide array of HTTP clients. The natural client for our server application, of course, is the web browser. Each browser should be compatible with the HTTP specification, but the problem is that there are a great number of inconsistencies between the implementations of different browsers, which leads to many possible incompatibilities. The best thing to do is to support the smallest number of HTTP features that will work with the broadest number of client browsers. This requires delving into the specification itself and then trying out the implementation with a number of different browsers to see what is required for minimal interoperability, then moving forward with any additional features required by the application itself.

Another thing we can do is look at the target client. If our application is going to have an HTTP server that provides a simple web interface with authentication enabled, we may be able to take a look at the form of the authentication requests and responses accepted and sent by a particular web browser. Assuming that we can get away with only supporting one browser, we can actually look at the messages sent by the browser and eliminate some of the parsing on the server end to reduce code size. We can provide robustness by simply checking for a few common properties of the messages received by the server device. If the server does not recognize something, then the connected client cannot be supported, and an error message will be sent back to the client.

Security for server-side HTTP can include a number of different protocols and options. At the very basic level, we can implement digest authentication using a standard hash algorithm such as SHA-1. For a little bit better security, we can add in some form of encryption using AES or some other symmetric-key algorithm. Obviously this would require shared keys, but depending on the application this may be adequate. If we had the advantage of the device being on a VPN or other such hardware-encrypted channel, this would satisfy our security requirements, and a simple password authentication (no digest/hashing required) could suffice for authentication purposes. If we had a bare device out on a public network, however, we would need a more robust security setup. For HTTP, this means HTTPS—HTTP secured using SSL.

We will not go into the details of SSL right now (we have an entire chapter dedicated to the protocol), but HTTPS is a fully specified protocol (RFC 2818, "HTTP over TLS") in which HTTP is used over a socket-level implementation of SSL (or TLS—SSL's successor). Something to note about HTTPS: in theory, HTTPS is simply HTTP plain-text used over a secure SSL channel (socket). In practice, there are a couple of situations that need to be handled. First, HTTP is an inherently stateless protocol, and SSL requires the state to check for attacks (especially truncation attacks). HTTP normally closes a connection by simply closing the TCP/IP socket. This makes life difficult for SSL, which requires knowledge of the socket being closed, or an attack/error condition is met, indicating to the SSL implementation that an unauthorized close of the socket occurred. To handle this, an HTTPS server should try to initiate the connection close sequence with the client so that the socket can be closed without signalling an unauthorized termination of the channel.

From the client side connecting to the HTTPS server, the most notable difference between secure HTTP and insecure HTTP is the addition of a certificate provided by the server (more information on SSL certificates in the chapter on SSL) and redirection to TCP port 443. Port 443 is reserved for HTTPS in the same manner that port 80 is reserved for plain

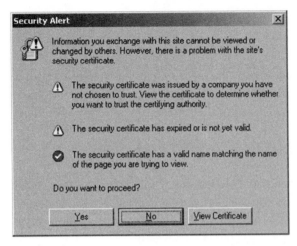

Figure 3: Screenshot of Internet Explorer Receiving a Possibly Invalid Certificate

HTTP. All modern web browsers support some form of HTTPS, and the port redirect (which is triggered by prefixing the URL with "https" in place of "http") usually happens without any user interaction. If the certificate is known by the browser, or has been authenticated by one of the browser's affiliated Certificate Authorities, then the browser simply loads the secure page. If the certificate is unknown, malformed, or is not independently authenticated, then the browser will display warnings prompting the user to decide whether or not to proceed with loading the page, as seen in Figure 3. Every browser has a slightly different way of notifying the user about errors (which, if you remember, are possibly attacks), but those behaviors are somewhat dictated by SSL and are generally similar. In any case, an error in the browser can be disquieting to a user, so it is usually best to handle these cases either through documentation or appropriate use of SSL certificates (again, see the chapter on SSL).

HTTP Client Web Interfaces

We just discussed server-side HTTP web interfaces, but what about client-side? Implementing client-side HTTP has some distinct advantages over server-side, especially when it comes to utilizing the built-in security mechanisms of the HTTP protocol. With the embedded device utilizing a client-side implementation, then the cryptographic random numbers used for digest authentication (the challenge numbers) are likely generated using more powerful hardware (on the server end), with much more capability of generating appropriately secure numbers. Besides this useful property, the client-side implementation can be much simpler than the server. It is much more likely that you will only need to support a single server with your client implementation than being able to support only a single client implementation with an embedded HTTP server.

The security of an HTTP connection is typically dictated by the server, so the security needs of an embedded HTTP client are dependent upon the specific application. If the client will only connect with a single server, then the security can be tailored to match that server's needs, allowing for a reduction in application code size, since the security options need not be comprehensive. If the client will be contacting multiple, possibly unknown servers, then the scope of security mechanisms supported should be broad and flexible.

Combination Client/Server HTTP Applications

A combination approach may be desirable if the embedded devices in the application need to communicate with one another, as well as with larger PC or server machines. The disadvantage here is that we need to support a broader range of technologies in order to support both client and server HTTP, which will require additional space and memory.

HTTP provides a slick, usually highly graphical interface for many applications. However, sometimes the application does not need (or simply cannot support) a full web server. As a result, a console interface may be more appropriate. In the next section, we will discuss console interfaces and a couple of options that may be used to secure them.

Console Applications

In many cases, the embedded device on the other end of your network may not be sophisticated enough to bother trying to implement a full HTTP stack or go all out with full-blown SSL and HTTPS. Of the alternatives to HTTP, one of the most common methods for communicating that is still in widespread use today is the venerable Telnet. The Telnet protocol is exceedingly simplistic, providing basic text communication capability that is designed explicitly for implementing remote console interfaces. Telnet is a natural technology for embedded devices since it has very little overhead and can be implemented very simply. Add to that the fact that just about every PC under the sun supports some form of Telnet client—there are literally hundreds of client applications available.

Telnet, with its low overhead and simple implementation, is great for embedded applications but suffers from one major drawback: it has no security options to speak of. Like HTTP, Telnet is designed to send information in the clear. If you are asked for a password over Telnet, you should know that if you type it in, you will be sending a clear-text copy of that password over the network. Telnet is so easy to use and so horribly insecure that many organizations ban its use outright, and block all Telnet connections and ports.

So you are now probably thinking that Telnet, albeit very attractive for embedded development, is too much of a liability to use. This is not necessarily true. While the use of Telnet by itself over a public network is probably not a good idea, it can still be useful if you have a VPN setup (using IPSEC, for example) between you and a private network to which the end device is connected. Obviously, if the network is private and controlled, plain Telnet can be great.

So it is wonderful that we can use plain Telnet if the network is secure, but there are many times when that is not the case. What then? Well, it turns out that there are several alternatives to plain Telnet that provide complete all-around security. The most recognizable of these alternatives is the Secure Shell protocol, or SSH. There is sometimes confusion about the relationship between SSH and the Secure Sockets Layer (SSL), since the abbreviations are very similar and they sometimes appear to function the same way. It is true that you can implement Telnet over an SSL connection, but that arrangement is typically referred to as Secure Telnet. In reality SSH is a complete replacement for Telnet and has basically nothing to do with SSL. It also provides a number of additional features that make it more of a protocol suite than a single protocol. That being said, SSH and SSL share a lot of properties: They both utilize public-key cryptography to exchange symmetric encryption keys, they both utilize a similar handshake method to establish a connection, and they both have similar integrity checking mechanisms using various hashing schemes.

SSH is fairly popular, being that it is a simple, efficient, secure implementation that can be used in place of more general solutions like SSL. This specialization may also prove to be the downfall of SSH, as it may be replaced by SSL-based solutions in systems that already need SSL for something else—why reinvent the wheel when you have a general-purpose protocol at hand?[1] We will not cover the details of the SSH protocol in this book, since we spent an entire chapter on SSL, and the concepts behind the two protocols are very similar. After reading Chapter 4 on SSL, it should be a fairly easy transition to understanding SSH. In reality, both SSH and SSL-based secure Telnet are nearly equivalent from the user's standpoint. The protocols are designed to be mostly invisible (SSL more so than SSH), so you can build your console to be compatible with either. Obviously, if you already need SSL for something else in the system, it would not be very efficient to bring in an entirely different protocol stack (unless you *really* needed SSH for some reason). In any case, it is a good idea to use one or the other, since Telnet by itself just is a bad idea for most applications.

So just what would secure Telnet or SSH be good for? Well, many applications may need some type of remote interface for basic configuration and maintenance, but do not need to have a full-blown Web-based interface or custom executable. Telnet uses plaintext only, no fancy graphics or fonts, so it is fast and has the added benefit of not requiring serious translation for the device to understand. Given this, what does an SSH application look

---

[1] One thing that SSH does have going for it is that it comes prebundled with a number of Unix and Linux distributions—in particular, SSH is quite useful for IT management for remote servers.

like? Well, like we discussed above, the protocols are designed to be mostly invisible to the user, and that pretty much applies to the application designer as well. Implementing Telnet over SSL is as simple as replacing any TCP/IP socket calls with the appropriate SSL API calls in the Telnet source code. With SSH, the basic implementation should be very similar to Telnet itself (providing basic console functionality), depending on the implementation.

Console interfaces are quite useful for remote configuration, since they provide low over-head and there are numerous client programs available for virtually all operating systems. There really is no way to implement a simpler interface, so the use of a secure console-based implementation is a natural fit for embedded systems. In the next section, we will start to look at a specific variant of console protocol that is specifically used to transfer files between systems over a network. As we will see, the File Transfer Protocol (FTP) and its variants can provide useful functionality, but like anything else that uses a network, can benefit from security.

File Transfer Protocol

The File Transfer Protocol (FTP) has been around since the early days of networking. The venerable protocol provides one of the simplest functions a computer can perform—copying and moving files from one location to another. FTP provides a network interface for file transfer operations, so that files may be shared between physically remote systems. The typical interface to FTP is a console, similar to Telnet or other remote console protocols, but the interface is just a wrapper for the actual protocol. Like other older protocols, however, FTP suffers from a complete lack of real security. Sure, an FTP client may prompt the user for a username and password, but the protocol itself simply transmits everything in plaintext over the network. In this section, we will look at FTP and some of its variants, as well as ways to secure those protocols.

FTP can be used for a wide range of functions in an embedded system, from data gathering to uploading firmware updates. However, FTP by itself suffers from various security flaws that make using it stand-alone somewhat hazardous. In RFC 2577, the authors (of the RFC) outline several security issues that should be addressed in any FTP implementation, the most problematic of which are a so-called "bounce attack," unlimited password attempts, and sending information (including passwords) as plaintext. The bounce attack allows an attacker to use an FTP proxy server to forge or disguise the origin of attacks on well-known TCP ports by re-routing commands through an intermediate machine (the FTP proxy), as seen in Figure 4. This attack would be problematic for an embedded application only if the FTP proxy service was enabled, but it would allow for the device to be used as the interme-diate proxy for the attack. The other vulnerability would arise from an intermediate proxy attacking the device itself, but that problem could be mitigated through the use of appropri-ate authentication methods. The reason this vulnerability exists is that FTP supports sending

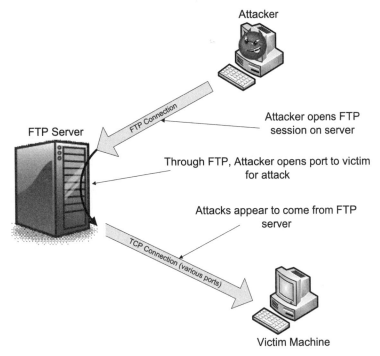

Figure 4: FTP Bounce Attack

files to different TCP ports. The FTP protocol will happily forward a file to any port the user chooses, and if that file contains relevant commands for a particular service in ordinary text, the file will be treated as incoming commands. If the FTP server and the victim machine are behind a firewall, it should be easy to see that an attacker could access the victim machine even if it was not directly accessible from outside the firewall. A couple of ways to secure against this vulnerability are to restrict the ports the FTP server can use to unused TCP ports, or to simply disallow the target port to be redirected.

Somewhat more problematic for embedded applications are the issues with passwords in FTP. The FTP spec allows for unlimited password attempts after a failure. This "feature" enables brute-force password guessing through automated means—in an embedded application with limited resources, the bigger threat would really be a denial-of-service condition while the attacker continually tries new passwords without letting other client connections through. The limited resources would likely not respond quickly enough for the brute-force attack to be as effective as it would be if the server was a more powerful machine.

Finally, the plaintext transmission of password information is exactly the same issue we discussed as part of HTTP authentication and the Telnet protocol. The only real solution here is to use FTP over some form of security protocol such as SSL or IPSEC. Using FTP over SSL is very similar to the use of HTTP over SSL or TLS. Described in RFC 4217, securing FTP over TLS is fairly straightforward. There is also a version of FTP that runs over SSH.

The full FTP protocol offers a large number of options for implementation, many of which are not strictly necessary for some applications. For this reason, the Trivial File Transfer Protocol (TFTP) was developed. Originally described in RFC 783, TFTP was designed to be the simplest file transfer protocol available. It was only intended for the reading and writing of files, without any of the other features (such as listing directory contents) of FTP. However, a flaw in the design led to an error condition known as "Sorcerer's Apprentice Syndrome" (or SAS, named after the animation sequence from Disney's classic movie *Fantasia*). The problem was that a timeout mechanism was used to determine if a packet needed to be retransmitted. TFTP was implemented over UDP, so the timeout was necessary to assure that the entire file was transmitted, but it hid an interesting problem: If certain packets were not actually lost but simply delayed enough to violate the timeout (usually due to network traffic congestion), the packets would be considered lost. In TFTP, the receiving implementation sends a message (ACK) back to the sender after each packet is received. If the ACK is delayed, the sender may send a duplicate packet. Upon receiving the duplicate packet, the receiver would send a second ACK. Each time the sender receives an ACK, it transmits the next packet in the current sequence, so the duplicate ACKs result in duplicate packets. The additional overhead leads to more congestion and the result is a flurry of duplicate transmissions which can slow the network to a crawl (see Figure 5). The fix for the problem involves preventing the sender from transmitting a packet after the first ACK is received for that packet (ignore duplicates). The SAS problem and solution are described in RFC 1123. A revision of TFTP with the fix is described in RFC 1350.

The above discussion about the SAS problem with TFTP illustrates an inherent problem in security—you may implement a protocol exactly to spec, but the protocol itself may be flawed. For this reason, it is important to keep up with the latest updates to any standard protocols you may use in an application.

Having discussed the issue with the original TFTP protocol, we will now look at TFTP itself. TFTP is of particular interest to embedded systems designers, since it can be used as a method for updating firmware. Implemented as part of a BIOS or low-level system, TFTP could be used to retrieve an application binary image to be executed on the remote device. Obviously, this simple protocol provides no security from tampering or eavesdropping, so something needs to be done to prevent attacks on the device. One possibility for securing TFTP would be to use it with SSL, but that really hampers the concept of a simple system used for bootstrapping the device. What would be more practical would be to use a cryptographic algorithm, such as AES, to encrypt the file before transferring. If simplicity is necessary then even AES may be too complicated, in which case RC4 may be a better

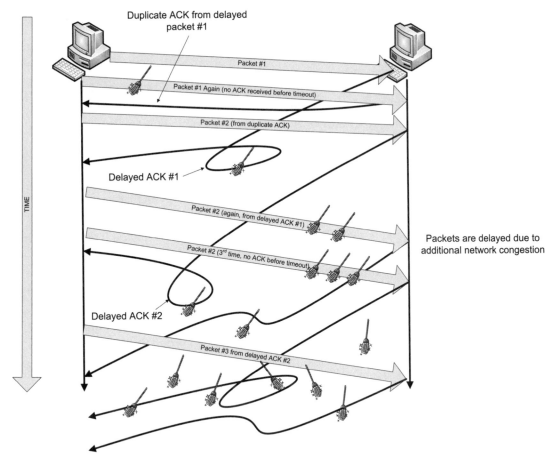

Figure 5: Sorcerer's Apprentice Syndrome and TFTP

choice. Even simpler would be to use a hash of the program to verify its integrity (this obviously does not prevent eavesdropping). If true encryption is required, it means a key would need to be stored on the device. This is a security hazard by itself, but for now we can assume that the device (see Chapters 9 and 10 for hardware security and issues with physical security).

To secure a firmware upload using something like RC4, the device would have to be initialized with some key (and an initialization vector since RC4 is a stream cipher). It is recommended that a protocol be implemented that allows this key to be changed upon deployment, so the compromise of a single device would not lead to a possible compromise of all other identical devices in the field. Even better would be to initialize every device with a unique key (the logistics of this may be prohibitive, however). Having the key, any file to be uploaded to the device would first be encrypted and then TFTP would be used to send the encrypted image. To further add to the security of the image, a checksum should be calculated on the original image and included in the encrypted payload. Upon receiving the encrypted payload, the device would immediately decrypt the file and load it into memory.

Once the program image has been received by the device, a check should be done on the file to make sure it is valid before overwriting any existing program. To do this, some identifying sequence (basically a password) should be the first thing encrypted and sent; the device would then look for that sequence before accepting any further packets. This presents an interesting problem for resource-constrained systems, since there may not be enough memory to store a complete image on the device. In this case, the image could be split into smaller chunks and loaded piecemeal. Another possibility is that the loading firmware has a mechanism that prevents a new image from executing until some check is done or encrypted command is received. In any case, the resident firmware that handles the loading of new images should be written to protect against the execution of arbitrary program images without some form of integrity check.

Our TFTP uploading example above uses a combination of authentication (using the encrypted "password"), protection against eavesdropping (using encryption), and integrity (from the checksum) that will thwart most attackers. That being said, this simple example is not sufficient to protect against all threats. In Chapter 11, we will look at a similar protocol for securing commands sent to a PIC-based application. The PIC case study covers the concept of a relatively secure protocol in much more depth than our simple example with TFTP. In any case, a secure TFTP boot system based on the concepts described above may be a very useful tool for firmware updates (or any other type of file transfer) that would not be too much of a burden on resources.

We have so far discussed FTP and TFTP, but there are literally dozens of other file transfer protocols in existence—far too many to cover here. Before we continue, however, there is one other file transfer method that may be of interest: HTTP uploading. Originally described in RFC 1867 as an extension to HTML forms, uploading files through an HTTP interface can be quite useful. In an embedded application, HTTP upload could allow an end user to update web pages or firmware through a simple and intuitive web-based interface. Securing HTTP upload is the same as securing HTTP, so we will not go into the details here, but as long as HTTPS is implemented, HTTP upload will be secure. In Chapter 12, we cover an application using the Rabbit microprocessor with an HTTPS server, and although the case study does not use HTTP uploading, it could easily be updated to without affecting the rest of the application.

In the next section, we will look at several other widely-used protocols, how they apply to embedded systems, and what types of security considerations need to be made when using them.

Email, DNS, DHCP, and SNMP

To close out the chapter, we will briefly touch upon a few other protocols that will be of interest to many embedded systems developers. Though diverse in functionality, these protocols all share a common property: They are distributed throughout a network, rather than being simply available between two devices. In the most trivial cases, each of these protocols could be used between two devices, but that is not the most common deployment for them. The protocols discussed previously in this chapter can also be deployed as distributed applications, but are often thought of in terms of a simple single-client/single-server environment, especially when dealing with individual embedded devices. In this section, we will look at four of the most common and widely used distributed protocols, Email (or the Simple Mail Transfer Protocol, SMTP), the Domain Name Service (DNS), the Dynamic Host Configuration Protocol (DHCP), and the Simple Network Management Protocol (SNMP).

The term "email" is normally associated with the Simple Mail Transfer Protocol, or SMTP, which is described in RFC 2821 (this is an update of the original SMTP that was described in RFC 821). Probably the most recognizable and common form of communication on the Internet, email suffers from a complete lack of security in its default state. This is one of the reasons that we now have numerous issues with spam and worms. In the embedded world, email provides a simple and effective form of notification when an embedded device requires attention. SMTP can be easily implemented, and email addresses are easily set up. With practically everyone having at least one email address, it makes sense to use SMTP for communications that do not require low latency (email can take a while to move through a network). The complete lack of security, however, can make email a big liability. First of all, any device that is configured to receive email can be the target of an "email bomb," where thousands of repeat emails are sent to a single address in one large burst, effectively cutting off any communication with the device (if not crashing it altogether).[2] The other major problem with email is that there is no privacy whatsoever—all messages are sent in plaintext. For many years, there have been solutions to protect emails in transit, such as the original implementation of PGP, but none have really caught on. If you plan to use email in an embedded application, there are a couple of things to think about: first, you need to know if the email needs to be encrypted to protect against eavesdropping. The easiest thing to do for encryption is to simply encrypt the email payload (the message) and then send it as a file via email. As long as the encryption does not suffer from common vulnerabilities (use checksums and counters to make every message unique and update the keys regularly),

---

[2] The term "email bomb" may also refer to an email containing a payload (usually an attachment) that launches a virus or worm when opened or executed. Sometimes the emails themselves can be designed to exploit vulnerabilities in email clients.

this scheme should work pretty well. However, even if the payload is encrypted and cannot be looked at, there is no guarantee that the email will actually arrive at the correct address. It would be possible to send a return email indicating that the message had been received, but you would be required to implement that in your application. It would be better to simply use a protocol like SSL that guarantees the messages are received on the other end. The primary issue with email as a communications medium for embedded applications is that it relies on the network to deliver the email. Lower-level protocols like TCP also rely on the network, but at a lower level. Email servers can cache emails or choose to filter them out, whereas a TCP connection will either work or fail immediately. If the possibility of losing a message is not a big deal, but having the contents read by an attacker is, then email with encrypted payloads would be fine. An example application that would use this scheme would be a home medical monitoring system that records noncritical information (such as average blood pressure) and sends it off once a day to a doctor. The information, not being of a time critical nature, can be delayed or perhaps even lost, but it would be important to keep that information private. In any case, email is useful for particular applications, but is probably not the best all-around communications mechanism for most systems.

Email is an example of a communications protocol that utilizes various servers to route messages through the Internet, which requires a mapping of email addresses to particular systems. In a similar fashion, the Domain Name System, or DNS, relies on a distributed network of servers to provide a lookup service that maps IP addresses to human-readable names, called domain names or Uniform Resource Locators (URLs). DNS is a complex combination of protocols and functionality that provide a dynamic mapping of network addresses, allowing for multiple machines to be represented by single domain name. Generally speaking, most embedded applications of the scope we are discussing will not really deal with DNS too often, but we mention it here because of its importance to networking in general. One security issue to be aware of with DNS would be "cache poisoning," where an attacker changes a DNS mapping in a DNS server, causing all requests for a particular name to be routed to a machine of the attacker's choosing. If an embedded device is referred to by a URL instead of an IP address, or if the embedded device needs to access another device using a URL, cache poisoning may be an issue. Unless the device itself will provide DNS services, there is not much that can be done, other than provide appropriate authentication and data protection so that if requests are rerouted at the DNS level, the effects will be minimal.

DNS may not be generally applicable to many embedded applications, but there is a case where having a URL refer to a specific device may be desirable. If the device in question does not have a specific IP address, but instead relies on a dynamic addressing scheme, then having a URL would be useful if a user needed to access the device (the other option would be to use a discovery protocol to learn the address of the device when that address changed). One protocol that is commonly used to provide IP addresses in a dynamic fashion is the Dynamic Host Configuration Protocol, or DHCP. The latest incarnation of DHCP is described in RFC 2131, but a large number of documents flesh out the nuances of the protocol. For embedded devices, DHCP is only important in that it is commonly used to assign addresses to devices to avoid collisions between static IP addresses. The DCHP specification does not describe any mechanism for securing DHCP, and notes that the protocol is completely insecure. DHCP is implemented as a client-server application, and embedded devices would normally be DHCP clients. The DHCP server would be a system or device responsible for assigning addresses. Like DNS, DHCP is really outside the scope of this book since it only peripherally applies to most embedded applications at the level we are discussing. It should be sufficient to protect against DHCP vulnerabilities by implementing appropriate authentication, privacy protection, and integrity checking for all communications implemented on the device.

Finally, and to close out the chapter, we will briefly discuss the Simple Network Management Protocol, or SNMP. The workings of SNMP are far beyond the scope of this book, but it is worth mentioning because it provides functionality that can be very useful for embedded applications. Originally designed to provide monitoring and configuration services for remote servers, SNMP has gone through three incarnations, referred to as SNMPv1, SNMPv2, and SNMPv[3]. All three variants of the protocol are incompatible, but provide similar services with each subsequent version improving on its predecessor. Generally speaking, SNMP versions 1 (RFC 1157) and 2 (RFC 1441) are considered obsolete, but are still useful and are widely implemented. SNMP version 3 is described in several RFCs (3411–3415). In general, SNMP is anything but simple, so securing it is also a complex task. For this reason, we will leave the discussion of securing SNMP to other references. If you are interested in learning more about SNMP and its associated security issues, a good place to start is with the RFCs, but there are a large number of web references and books on the subject as well. For a general overview of the security issues with SNMP, another good place to look is at the CERT[3] website (www.cert.org). CERT calls out SNMP as having some major security issues, since it does not implement even the most basic forms of protection and authentication.

Wrap-Up

This chapter has been all about the Application-Layer and Client/Server protocols that provide the bulk of communications functionality that developers see when developing

---

[3] CERT (not an acronym) is an organization that keeps tabs on security issues pertaining to the Internet. CERT also has an emergency response team for Internet security issues, and is located at Carnegie Mellon University in Pittsburg, PA.

applications. We have covered some of the most common and popular protocols and tried to give some insight into how they relate to embedded systems security, but this has by no means been a complete treatment of any of the protocols mentioned. It has been provided primarily as a springboard for research into these protocols as they apply to specific applications. There are literally hundreds of protocols in use (take a look at the reserved TCP port list sometime) today, and some may be more applicable to your application than others.

In the next chapter, we will look at how protocols and algorithms can be optimized for embedded applications with limited resources. There are a lot of caveats when optimizing anything that, if ignored, can lead to serious security vulnerabilities. We covered some of the topic in Chapter 5 when we discussed embedded security, but Chapter 8 focuses on optimizations that are safe and outlines what should be avoided. For situations that require higher performance, Chapter 9 will look at hardware-based solutions that provide optimal performance without having to optimize code.

Choosing and Optimizing Cryptographic Algorithms for Resource-Constrained Systems

In this chapter we will finally look at cryptographic algorithms and how to make them work for embedded machines. Cryptography presents the biggest challenge to any security solution to be implemented for an embedded device because of the requirements of these computationally complex algorithms. Cryptography is notoriously expensive when it comes to clock cycles, and many algorithms are not too friendly to smaller memory devices. We will look at various strategies for choosing algorithms for specific applications, and look at some specific algorithms as well as some strategies to avoid some of the more expensive cryptographic operations without compromising security.

First of all, we will look at whether cryptography is even necessary. Some applications can actually get away without using traditional cryptography. These applications utilize other mechanisms, such as hashing algorithms, in order to provide some assurance about data. The big advantage here is that hashing algorithms are typically many orders of magnitude faster than symmetric or public-key cryptography. One classification that will help us make the decision is whether or not we care about eavesdropping. If we only care that the data is reliable, and do not care who gets it, we can definitely avoid cryptography. We will look at these types of applications and come up with some other general rules for determining if an application requires the more expensive operations, or can get away with less.

We will look at cryptographic hashing, and discuss some optimizations and other tricks that can be used to speed up commonly used algorithms. Due to recent discoveries of weaknesses in the commonly used algorithms MD5 and SHA-1, we will also look at the future of hashing, and see if there is an alternative method, or new algorithms we can choose.

For those applications that require absolute secrecy, there are many options that will help the embedded developer meet the goals of performance, cost, and security. We will look at some specific algorithms, such as DES/3DES, which is slow and obsolete but still necessary for some applications. AES, the replacement for DES will also be covered, and specifically, some of the hardware options available. Additionally, we will look at RC4 and other ciphers that provide a marked performance advantage over the bulkier DES and AES, but are not as provably secure.

Finally, we will cover public-key algorithms, and specifically, RSA, by far the most commonly used public-key cipher. Unfortunately, RSA and other public-key algorithms are extremely slow, requiring hardware assistance on many platforms—not just small embedded systems. However, these algorithms are essential to certain protocols (primarily SSL and SSH), so we will look at ways of handling the computationally intense public-key operations, including software tricks and hardware assistance.

Do We Need Cryptography?

One of the first steps in building a secure embedded system is to see if cryptography is actually needed. Whenever security is discussed, many engineers will immediately think of cryptography as *the* solution, when in fact, many options may exist that do not strictly require cryptography. Sure, cryptography is an important part of securing applications, but it is not a security panacea, nor is it always necessary for building a secure system. To many people, cryptography is a mysterious and confusing, almost magical area of computer science that is reserved for cloistered geniuses at top academic institutions and the US National Security Agency. The NSA is mysterious and secretive, and it is believed to employ some of the greatest cryptographers in the world, only adding to the idea that cryptography is an untouchable discipline.

There have been some attempts over the last couple decades to bring cryptography out from behind its cloak of mystery, most notably with the publication of *Applied Cryptography* by renowned security expert Bruce Schneier. Schneier, more than any other individual, has brought cryptography into mainstream computer science by unveiling the techniques employed in most common cryptographic algorithms. That being said, a stream of inaccurate portrayals of cryptography in Hollywood and television combined with unrelenting advertising campaigns for various security products have served to only keep those not involved directly in computer science in the dark. For this reason, it can be difficult to explain to a managers or hardware engineers what exactly cryptography is, let alone convince them that it will not solve their security problems.

Given the confusion surrounding cryptography, it can be extremely difficult for a systems engineer to determine what type of cryptography, if any, is needed. The truth is that in some circumstances, no cryptography is needed at all. For instance, if you do not care about anyone seeing the information being transmitted, but rather you are only concerned that it is not tampered with in transit, you really do not need a complete cryptographic suite to suit your needs. For example, let's take a look at publicly available stock market reports. For many investors, keeping tabs on the latest price of a stock can mean thousands or millions of dollars being made or lost. The price of stock in a public company is readily available, but if someone were able to change that price in transit during a trading session, that person would be able to wreak havoc upon unknowing investors. Obviously, there is a security risk for any system that transmits this data, but we do not need to hide it from anyone (unless your goal is to provide only your customers with the most up-to-date information, but that is another matter). Instead, we can just use a simple hash algorithm to verify the integrity of the data being transported. Not being a full-blown encryption scheme, the system's requirements can be lowered to support just the hashing (or those resources could be used for improving the performance of other parts of the system).

In order to see what type of security an application will need, we can divide applications into several distinct categories based upon the type of information each application deals with. After all, the purpose of computer security is to protect information, so this is a good place to start. The following categories should be sufficient for most applications, ordered from the lowest level of security to the highest level of security:

- No security required (may be optional)—applications such as streaming video, noncritical data monitoring, and applications without networking in a controlled environment.

- Low-level security (hashing only, plaintext data)—applications delivering publicly available information such as stock market data, data monitoring, networked applications in a controlled environment, and applications requiring an extra level of robustness without concern about eavesdropping. Another example of this would be the distribution of (usually open-source) source code or executables with an accompanying hash value for verifying integrity, usually an MD5 or SHA-1 hash of the files that is to be calculated by the end user and compared to the provided hash.

- Low-medium security (hashing plus authentication)—general industrial control, some web sites, and internal corporate network communications.

- Medium-level security (cryptography such as RC4 with authentication, small key sizes)—applications dealing with general corporate information, important data monitoring, and industrial control in uncontrolled environments.

- Medium-high security (SSL with medium key sizes, RSA and AES, VPNs)—applications dealing with e-commerce transactions, important corporate information, and critical data monitoring.

- High-level security (SSL with maximum-sized keys, VPNs, guards with guns)—applications dealing with important financial information, noncritical military communications, medical data, and critical corporate information.

- Critical-level security (physical isolation, maximum-size keys, dedicated communications systems, one-time pads, guards with really big guns)—used to secure information such as nuclear launch codes, root Certificate Authority private keys (hopefully!), critical financial data, and critical military communications.

- Absolute security (one-time pads, possibly quantum cryptography, guards with an entire military behind them)—used to secure information including Cold War communications between the United States and the Soviet Union regarding nuclear weapons, the existence of UFOs, the recipe for Coca-Cola, and the meaning of life.

The above categories are intended as a rule-of-thumb, not as strict guidelines for what level of security you will want for a particular application. Obviously, your particular application may fit into one of the above categories and require a higher level of security than indicated. You could also reduce the recommended level of security, but remember that security levels are always decreasing as hardware improves—what we consider to be "high" security today may not be very secure at all in 10 years. The prime example of this is the fact that 56-bit DES was considered nearly "unbreakable" when it was first introduced, but was broken just 20 or so years later with a concerted effort. Now, more than 10 years after that, DES is practically obsolete, and even the beefed-up triple-DES is showing its age.

If you haven't gathered by this point, this book is primarily focused on applications requiring security up to medium level and some medium-high level (as described in the categories above). If your application falls into the medium-high category or higher, you will definitely want to do some serious research before jumping into implementation. In fact, you should do a lot of research if your application needs security at all. This book and others will be useful to determine what you need for your application, but you shouldn't ever take one person's or book's viewpoint as gospel. It is highly likely (pretty much guaranteed) that any single viewpoint on security will miss something important that can come back to haunt you later. That being said, we will move forward and attempt to make some sense of cryptography for systems not strictly designed for it.

Hashing—Low Security, High Performance

Ideally, we would not need any security in our applications at all, instead having the luxury to focus all resources on performance. Unfortunately, this is not reality and we have to balance security with performance. Fortunately, the security-performance tradeoff is relatively linear—higher security means some performance loss, but the balance can be adjusted fairly easily by choosing different algorithms, methods, and key sizes. For the best performance with some level of security, you can't beat the hash algorithms. As far as resources are concerned, they use far less than other types of cryptographic algorithms. Not strictly cryptography in the classic sense, hash algorithms provide integrity without protection from eavesdropping—in many applications this is all the security that is really needed.

How do we actually make hashing work for us in our applications? Probably the most common use of hashing is to provide a guarantee of integrity. The concept behind cryptographically secure hashing is that hash collisions are very difficult to create. Therefore, if you hash some information and compare it to a hash of that information provided to you by a trusted source and the hashes match, you have a fairly high certainty that the information has not changed or been tampered with in transit.

Another mechanism that employs hashing to provide a level of security is called "challenge-response," after the way the mechanism works. In a challenge-response operation, the system or user requesting the information provides a "challenge" to the sender of the requested information. The sender must then provide an appropriate "response" to the requestor, or the information is deemed invalid. There are many variants of this simple method for verifying the authenticity of information, but for our purposes, the variant we will talk about uses hashing to prove that the requestor and sender (client and server) both know some shared secret only known to those systems. The challenge is typically some cryptographically secure random number that is difficult to predict, and it is sent by the client to the server (or from the server to the client, as is the case with some HTTP applications). Once the server receives the challenge, it calculates a hash of the challenge and the secret, and sends the hash back to the client. The client also performs the same operation on its copy of the secret, so when it receives the hash it will know that the server knows the same secret.

The reason the mechanism works is because the server could not produce the correct hash value unless it knew the secret. The random number for the challenge ensures that every hash sent back to the client is different; otherwise an attacker could simply send that value to the client and there would be no security. The challenge-response hashing described here does suffer from a man-in-the-middle vulnerability, since an attacker could intercept the message in transit from the server back to the client. This attack requires the attacker to have access to the routing system of the network in order to spoof the address of the server and the client so it can intercept and retransmit the messages. However, for security at this level, it can usually be assumed that breaking the network routing will be sufficiently difficult to provide a decent level of security.

So now we know how hashing can be used to provide different types of security, how do we actually use it in practice? There are several hashing algorithms available, but the most common are MD5 and SHA-1. Unfortunately, the MD5 algorithm has been shown to have some weaknesses, and there is a general feeling of uneasiness about the algorithm. SHA-1 has been the focus of some suspicion as well, but it has not had the same negative press received by MD5. In any case, both algorithms are still heavily used in the absence of anything better. For this reason, we will look at MD5 and SHA-1 for our examples. Though there is some risk that these particular algorithms will become obsolete, the methods described here for their use in cryptographic systems can be applied to any similarly structured hash algorithms.

In practice, many embedded development tool suites provide libraries for MD5, SHA-1, or both hashing algorithms. The algorithms themselves are not too difficult to implement (they actually consist of a number of quite repetitive operations), but the convenience of having a pre-tested implementation to use is quite nice. Hashing algorithms are fairly easy to optimize as well, so it is quite likely that the provided implementations will already be fairly optimal for your target hardware.

Using hash algorithms is quite easy, since they have only three basic operations that are provided in the user API:

1. Initialization, which sets up the state data structure used to actually perform the hash.

2. Hashing, which operates on the incoming data, typically in raw bytes (may also be text).

3. Finalization, which finishes up the hash, and copies the result into an output buffer.

In this book, we focus primarily on C-based development tools (being fairly universal in the embedded systems industry), so the following examples of hashing in practice directly apply to the C language, but the principles can be applied to other similar languages (such as Java or C++).

The basic operation of hashing algorithms uses a buffer, in C an array of *char* type, as a workspace for the hashing. The algorithms utilize a structure that provides state information across hashing operations, allowing for the hash to be added to in multiple operations, rather than all at once in a single hashing operation. This feature is the reason that each of the hashing APIs has three distinct operations. The algorithms are set up so that they can utilize the state structure to keep the in-progress hash intact without requiring separate support from the application designer. With this setup, once the user application is ready, it can provide an output buffer for the finalization operation. This is important for networked environments where several hashes may be in progress simultaneously—the user need only keep track of the hash state structures.

So what does hashing look like in a real application? In the following C program, the user input is hashed into an output buffer (actual SHA-1 API may vary):

```
#include <sha1.h>

#include <stdio.h>

main () {

        char input_buf[128], output_buf[20];

        struct SHA1_state sha_state;

        int i, input_len;


// Initialize the state structure, requires only a reference to the struct

SHA1_init(&sha_state);


// Get user input, make sure buffer is cleared first

memset(input_buf, 0, sizeof(input_buf));

scanf("%127s", input_buf);


// Hash the input, with a length equal to the size of the user input. Note that

// the input to the SHA1_hash function can be of any length

// !!! Danger, strlen can overflow, so we terminate the buffer for safety

input_buf[127] = 0;

input_len = strlen(input_buf);

    SHA1_hash(&sha_state, input_buf, input_len);


    // Finally, finalize the hash and copy it into a buffer and display

    SHA1_finish(&sha_state, output_buf);

    for(i = 0; i < 20; ++i) {

        printf("%X ", output_buf[i]);

    }

    printf("\n");

} // End program
```

Listing 8.1—Hashing with SHA-1

That's it—hashing is a very simple operation in code. Notice that we do some defensive programming when using the strlen function. Unfortunately, the C programming language does not have very good standard library support for protecting against buffer overflow.

In our little program example, if the user entered enough data to fill up the buffer to the end (more than 127 characters), we are relying on scanf to be sure that the last element of the array contains a null-terminator. In our program, the scanf "%s" type is used in the format string with the optional width format parameter, so it should not cause any issues for the call to strlen later. However, if someone was to change the scanf to some other form of input, then the assumption may be violated. For this reason, we add an unconditional null-terminator to the end of the array to be sure that strlen will terminate appropriately.

The hashing example above illustrates the use of a cryptographic algorithm to protect data, but it also highlights the fact that anything in the program can become a security issue. The use of standard C library functions such as strlen can lead to unexpected and unintentional behavior. Sometimes this behavior goes unnoticed; sometimes it leads to a crash. All it takes is one malicious attacker to find the flaw in your program and exploit it somehow. It may not be that the attacker gains access to the whole bank, but shutting down a few hundred automated teller machines could do a lot of financial damage. All that the attacker needs is for you, the developer, to stop paying attention. The example has a trick or two that help to keep the program safe, such as terminating the buffer (a simple operation that could be easily overlooked), but what if the algorithms themselves were the problem. In the next section we will look at some recent developments with 2 major hash algorithms (MD5 and SHA-1) that cast doubt on their continued utility.

Is Hashing Considered Dangerous?

In the past few years, cryptography has come into a lot of visibility, and the old faithful algorithms that have served us well for years are now being put to the test. The vast amount of information on the public Internet that needs to be protected has lead to a virtual stampede of corporations, governments, organizations, and individuals studying the security mechanisms that form the foundation of that protection. People on both sides of the law (and with varying levels of ethics) are racing to discover flaws in the most commonly used algorithms. After all, boatloads of money can be made if a security breach is discovered. For the "good" guys, the rewards are recognition and (hopefully) prompt fixing of the issue. The "bad" guys profit in numerous different ways. The end result is always the same, however: If an algorithm is broken, it usually means it's useless from that point on.

This insane virtual arms race has revealed that it is extremely hard to develop secure cryptographic algorithms (it's easy to write broken cryptographic algorithms), and it appears that hashing may be among the most difficult. The two major hash algorithms in use today (notably by SSL and TLS) are MD5 and SHA-1. At the time of the writing of this text, MD5 is considered "mostly broken" and SHA-1 is "sorta broken." What does that mean?

Well, there are various ways a hash algorithm could be broken from a cryptographic stand-point. Some of these are:

- Take two arbitrary but different messages and hash them. If you can easily calculate a hash value that is the same for these different messages (a "hash collision"), then the algorithm is somewhat broken, and potentially seriously broken.

- Given a hash value, compute an arbitrary message to hash to that value. If this is easy, then the algorithm is a little more broken, since this starts to get into the area where the flaw can be exploited.

- Generate a meaningful message that generates a hash collision with another mean-ingful message. If this is easy, then the algorithm is mostly broken, and it is highly likely it provides no security whatsoever. If this is true for an algorithm, it is very easy to fool someone into accepting incorrect information (or worse, damaging information such as a virus or Trojan horse).

Each of the above levels of compromise is based on the idea that performing these opera-tions on the algorithm is "hard" (infeasible given current technology and future technology for at least a few years). They all feed into one another as well, so if you can find an arbitrary hash collision, it is often easier to discover the other attacks.

Unfortunately, both MD5 and SHA-1 have been discovered to have vulnerabilities. For MD5 there are several demonstrations of ways to generate different meaningful messages that generate the same MD5 hash value. Granted, these operations are currently somewhat contrived and generally a little tricky, but it is only a matter of time until someone figures out how to do it fast and easy. Generally speaking, we don't need to rush out and pull MD5 out of all our applications, but if it isn't too difficult to do so, it is recommended. MD5 should not be used in new applications whenever possible.

The MD5 algorithm is fairly broken, but fortunately for us (and the rest of the world), SHA-1 is not as broken (yet). Researchers have discovered something akin to the first vulnerability (the arbitrary hash collision) in SHA-1, but as yet, there does not seem to be a way to translate that vulnerability into a security breach (as seems to be the case with MD5). Possibly by the time you read this, however, both SHA-1 and MD5 will be consid-ered obsolete and will be replaced by something new (or at least in the process of being replaced).

SHA-1 and MD5, albeit broken, are still in heavy use today and will continue to be for some time. They are so integrated into the security mechanisms that we have come to rely on that it will take years to pull them all out and replace them. Even then, many legacy systems may still require them in come capacity. This scenario obviously assumes there is a decent replacement. There are some contenders in the same family as SHA-1, but if that algorithm fails, it may be hard to tell if the SHA-1 vulnerabilities translate to its brethren.

One ray of hope, however, is that there may be another competition to develop a new cryptographic algorithm as was done with AES. Only this time, the target would be a hashing algorithm to replace the aging and ailing MD5 and the slightly less damaged SHA-1. Only time will tell what ends up happening with this. Heck, we might see a quantum computer implemented in the next few years that could make all of our current strategies obsolete overnight. We still need something in the meantime, however, and it worked for AES, so it may work for this too.

To Optimize or Not to Optimize · · ·

So far in this chapter we have focused on the choice of a single class of algorithms, hashes, as a way to get the most performance out of our security. While hashes are extremely useful for a wide variety of applications, they really cannot provide the same level of data protection that a "true" cryptographic algorithm, such as AES, can provide. One of the problems with hashes is that they produce a fixed-size output for arbitrary-length data. It doesn't take much thought to realize that if the message is larger than the size of the hash (and maybe even if it is smaller), some information is lost in processing. While hashes can work to give you a guarantee (large or small) that the data is intact and represents the original information, they cannot be used (at least directly) to encode the data such that it is extremely difficult for an attacker to get at it, but also can be decoded back into the original information given the proper "key." Hashes are, by nature, one-way operations, there is no way to build the original data from a hash value (in general, anyway—you may be able to guess at it if the message was small enough). To be able to effectively communicate information securely, it is vital that the data remains intact, albeit obfuscated, in the encrypted message. For this we need "true" cryptography as is found with symmetric and asymmetric encryption algorithms.

We talked about encryption algorithms and what they do in previous chapters, so we will not go into too much detail about symmetric and asymmetric cryptographic algorithms here, but the important things to remember are the essential properties of the different classes of algorithms. For your reference, we summarize the useful properties (and consequences of using) each of the classes of algorithms:

- Hashes—fast, efficient algorithms generally useful for verifying the integrity of data but provide no means to otherwise protect information.

- Symmetric—fast (relatively slow compared to hashes in general though), general-purpose encryption algorithms. The problem is that the keys need to be shared between receiver and sender somehow (and that sharing obviously cannot be done using a symmetric algorithm—keys must be shared physically or using another method).

- Asymmetric—slow, special-purpose algorithms that provide the useful ability to facilitate communications without having to share secret keys (public-keys obviously need to be shared, but that's generally pretty easy).

The rest of this chapter will cover the basics for optimizing cryptography for embedded applications. We covered hashing above because those algorithms are already very fast and can be used to provide a small level of security (assuming you don't care about eavesdroppers, anyway). The other cryptographic algorithms (symmetric and asymmetric) are much slower in general, and their relative performance may affect embedded applications significantly. With hashes out of the way, we can focus on the slower algorithms and just what we can do to make them a little faster.

Optimization Guidelines: What NOT to Optimize

Before we start any discussion about optimizing cryptographic algorithms, we need to cover a very important issue: what NOT to optimize. This is an extremely important concept, so it will be a dominant theme throughout this chapter (we even covered it to some extent in Chapter 5). The reason there are parts of these algorithms that cannot be optimized is that, in order to do their job correctly, they require some significant processing. If you go into optimization with your virtual machete blindly, you are more likely than not to remove something important. Optimization is not to be taken lightly, so the rule to remember is: "if you aren't sure about it, don't do it." A corollary to that rule is to always check with an expert if security is truly important—if you are just interested in keeping curious teenagers out of your system, then it may not be as important as, say, protecting credit card numbers.

So what are these things that should never be touched? Primarily, we want to be sure that any optimization or performance tweaking does not affect the proper functioning of the algorithm itself. Fortunately, this is usually easier to find than it sounds. Almost all cryptographic algorithms have test vectors that can be found to check basic functionality of implementations (be suspicious of algorithms that aren't easily tested). You can also use another implementation (Open Source software is always a good place to start since it is free and you can look at source code—just be sure to heed all the legalities inherent in using it). If you have broken the algorithm in your efforts to make it faster, it usually shows up quickly—one of the properties of cryptographic algorithms is that they are usually either broken or not, there are few cases where an implementation is "partially functional." The problems in implementations are usually with the handling of data once it has been decrypted (this is where buffer overflow and other issues come into play). The one case to watch for is if you are implementing the algorithm on both ends of the communication channel—you may break the algorithm, but since you are implementing both sides in the same fashion, you may not notice. It is always a good idea to check your algorithm against a known good (if you cannot, then it is a good sign you need to rethink what you are doing).

Another important rule for optimizing cryptography is that you should never try to tweak the algorithm itself. For one example, the AES algorithm uses a number of "rounds" to produce a secure encoding. A "round" is generally a sequence of actions that can be repeated any number of times, with the property that each additional round further increases the security of the resulting encoding (the concept of rounds is found in some symmetric-key algorithms such as DES and AES, but is missing in others, such as RC4). A typical AES mode such as 128-bit (key size) may use, for example, 18 rounds to encrypt a message. If you were to adjust your implementation to use fewer rounds, AES would still function and it would run faster. This may at first seem like a good idea, but you have compromised the security of your implementation—in fact, there are known attacks on AES when the number of rounds is reduced. Round reduction is one of the first things cryptanalysts look at when trying to break an algorithm, so you can be sure it is a bad idea to do it purposefully. Fortunately, if your implementation has to be compatible with other implementations, reducing the number of rounds will break the basic functionality of the algorithm. Again, the situation to watch for is when you are implementing both ends.

The example of round reduction in AES (or DES or any number of other algorithms) is just a specific case that can be used to demonstrate the more general rule. Don't try to change the design of a cryptographic algorithm (unless you are, in fact, a cryptographer). If the algorithm says that a certain property must be maintained or a certain feature must be present, it is there for a reason. That reason may be "the designer felt like it," but unless you know what you are doing, it is likely a bad idea to second-guess a cryptographer (remember Rule #2 from Chapter 2).

The third rule basically follows from the first two: cryptographic primitives need to be properly implemented or you might as well not even use cryptography. So what is a cryptographic primitive? Cryptographic primitives are the random number generators, entropy sources, and basic memory or math operations that are required by the cryptographic algorithms. For example, the Pseudo-Random Number Generator (PRNG) functions that generate random numbers from some seed value are extremely important to the security of your cryptography. You can have the strongest cryptography available, at even ridiculous levels (1024-bit AES? Why not? It may eventually be broken but for now it's pretty good), but if you always use the same predictable key value the implementation can be trivial to break. Case in point, the original implementation of the Secure Sockets Layer (SSL version 2) in the Netscape web browser was broken by a couple of grad students in a few hours. Did they find a way to attack RSA and DES? No, they simply looked at the random numbers that feed into the key generation in SSL (we cover this more in-depth in Chapter 12) and realized that the random numbers were being seeded with a very predictable source—the system date and time. Once they knew this, they could easily reduce the amount of work needed to break the encryption since the keys being used would always fall within a relatively predictable range. As a result of this, SSL version 2 actually became almost as insecure as a plaintext communication, and sent Netscape back to the drawing board (the fact that SSL version 2 is still in use today by some systems is baffling). The moral of the story is that you should always be sure that everything used by your cryptography is appropriately secure. Random numbers better be as close to unpredictable true randomness as you can get them—a physical source is usually best. For one solution, a combination of network traffic and external interrupts works fairly well. It isn't completely random, but it gets close enough for many applications (network traffic can be manipulated by an attacker to be more predictable, thereby making it less than an ideal source of randomness). Even better would be something based upon fluid dynamics or quantum physics. Unfortunately, however, the use of Lava Lamps for generating true random numbers was patented by Silicon Graphics in the 1990's.[1] If you do need truly random numbers, some manufacturers produce entropy-generating hardware, usually based on interference in electronic circuits. The author is a software engineer, so the principles behind this are a bit of a mystery, but if a physicist says its random, who can argue (besides another physicist)?

Now that we have expounded on the merits of having true randomness to make your system more secure, we will briefly cover the other parts of the third rule, such as making sure your PRNG is implemented correctly and your memory routines aren't doing something stupid. For the PRNG, it must have the property of not decreasing the entropy provided by the random seed. You could have a purely random seed, but if your PRNG always produced the same 10 numbers, it wouldn't be very effective. When implementing a PRNG, the same basic rules apply as when implementing a cryptographic algorithm. Don't change anything unless you really know what you are doing.

The final part of the third rule is to make sure that your memory and math operations are behaving properly. A perfect example of this is the use of a secure, on-chip memory to store a cryptographic key. If the chip is properly implemented, then that key should never leave that internal memory. As a programmer, however, you may have access to the key so that you can use it for your cryptographic operations. If you implement a copy routine that moves the key out into the primary system memory and leaves it there, you have essentially

---

[1] Patent #5,732,138: "Method for seeding a pseudo-random number generator with a cryptographic hash of a digitization of a chaotic system."

defeated the purpose of the internal memory. If you need to copy that key into system memory, make sure that you do it quickly and clear that memory as soon as you are done. It's not ideal (someone could access that memory while the key is there), but as long as there is a small window of opportunity, it is unlikely that an attacker would be looking and be able to extract the key.

To summarize our discussion of the basic rules of what not to optimize, we provide a brief description of the rules:

1. The implementation should not affect the basic functionality of the algorithm.

2. You should never try to optimize a cryptographic algorithm by reducing rounds or changing properties of the algorithm's design.

3. Obey basic cryptographic rules such as using an appropriate random source.

These rules are of course very generic, but should serve to protect you from most of the common mistakes that can be made when attempting to optimize cryptographic algorithms.

We have so far spent most of our time covering the basic ideas about what not to optimize, without providing much in the way of what actually *can* be optimized. The next section will cover some of the basics, providing a basis for any implementation effort. However, as we discuss the possibilities for optimization, keep the ideas from the above discussion in mind, since the line between what can be optimized and what should not be optimized can often be quite blurry. Most importantly, if you are not comfortable with the optimization (or even simply unsure), don't do it!

Optimization Guidelines: What Can We Optimize?

The basic rules of optimization apply to cryptography as they do in all other computer engineering disciplines—in other words, there are no rules! Unfortunately, optimization is somewhat cryptic (it is similar to cryptography in many ways) and often is considered "black magic" or "voodoo" by many engineers. While this is the case for some of the more esoteric optimization techniques (usually used on platforms employing some of the more advanced performance-enhancing features such as multistage pipelining, multithreading, instruction-level parallelism, and out-of-order execution), the optimizations we will discuss here are far more mundane, but can be extremely useful for certain architectures, especially those seen in the low-cost embedded hardware we are focusing on. For discussions on general optimization techniques, there are many excellent books written on the subject, but we are going to focus on simple tricks that are easy to spot and easy to implement.

The first thing to think about when looking to optimize a cryptographic algorithm is the math employed by the algorithm. Most cryptographic algorithms are very math-heavy, and utilize a number of mathematical primitives that may or may not be implemented in the most efficient manner possible. Two of the most common operations are the modulus and XOR, and though the latter is usually quite fast, the modulus can be quite expensive—especially if the hardware does not natively support division operations. The real trick in optimizing individual operations is to know what the compiler you are using will do to optimize the code, as well as what the underlying hardware is capable of. In some cases, it may be fine to simply let the compiler do the work for you, as would likely be the case with a multiple-thousand-dollar optimizing compiler from a company specializing in development tools, like something from Green Hills. However, with some of the less-expensive options, such as the GNU tools (gcc), C compilers for the Microchip PIC, or Dynamic C for the Rabbit microprocessors (see the case studies in Chapters 11 and 12 for more information), you may need to help the compiler out a little by doing some creative manipulation of the C code or even resorting to hand-coding in assembly language. If you are used to working with embedded hardware, assembly should not be too unfamiliar, but even if it is, there are usually an ample number of samples that can provide some insight into the types of things you can do. When dealing with limited resources, concerns about portability of the code take a back seat to efficiency and code size.

So what can we do to speed up these operations? If you are looking at an XOR operation, something implemented in hardware on most (all?) platforms, then there probably is not much you can do. If the data type is larger than the basic unit supported by the hardware (such as a 32-bit value on a machine with 16-bit integers), then look at what the compiler does. It is likely that the compiled code will be as good as anything you could write by hand. However, if the operation that you are looking at is a division or modulus operation, it may be (and is likely) that your compiler will utilize a more generic version of the operation than can be implemented in your particular situation. One trick employed by programmers who know the limitations of their tools is to look for powers of 2. Since we are dealing with binary machines (try finding a non-binary machine in production), anything that can be decomposed into powers of 2 (which is essentially everything) can benefit from some reworking of the code to utilize the inherent binary capabilities of the underlying hardware. Though this is by no means the only type of math trick we can use, it can be quite powerful when properly executed.

As an example of the power-of-2 type of optimization, let's take a look at one of the most common operations in cryptography—the modulus. A staple of cryptography, the modulus is used in algorithms from the simple RC4 to the more complex RSA. For our example, we will look at an operation that is specifically used in RC4, a modulus on an index into the internal RC4 lookup table:

```
char index;

int data;

…

// Derive new index from data

index = data % 256;
```

Listing 8.2—Power-of-2 Optimization Example Part 1

In the example, notice that the table size is a fixed 256 bytes. Conveniently (or by design), this happens to be a nice power of two (two to the eighth power, to be exact). This allows us to write the index calculation in a manner that may not be provided by the compiler itself (granted, this is a simplistic example using a constant that a good compiler should be able to do something with, but it is easy to explain). There are various ways to redo the expression, but we will look at an optimized solution utilizing a bit-mask. As it happens, a modulus by 256 will result in an integer in the range of 0 to 255, or FF in hexadecimal. Using what we know about binary, we can rewrite the modulus operation using FF as a mask:

```
char index;

int data;

…

// Derive new index from data using bit-mask in place of modulus

index = data & 0xFF;
```

Listing 8.3—Power-of-2 Optimization Example Part 2

The trick works because the remainder of any division by 256 will be represented by the lowest 8 bits (byte) of the result. This convenient property means that we do not care what is represented by the upper portions of the result since the result of the division will never have any bits set beyond the eighth bit (an astute reader will notice that we could simply assign the integer to the character and rely on the inherent truncation that occurs when assigning to a character in C, but it is usually a bad idea to rely on the language like that). As a result of the observation above, we can omit the division operation entirely. In an algorithm like RC4 where this operation or something similar is performed on *every byte* of the data being processed, the performance improvement can be staggering. In many cases, a function call to a complicated division routine will be replaced by one or two instructions, at a savings of potentially large numbers of clock cycles.

Generally speaking, anything that you can optimize that is related to the repetitive and mathematical nature of cryptography will be a boon for developing a secure but efficient application. In Chapter 9, we will look at some hardware options for performance enhancement, but for now we will stick with software solutions. The modulus example above is illustrative of a specialized trick that happens to work for a few different algorithms. Unrolling loops and other generic optimization techniques can also be quite effective, but as with anything there is usually a tradeoff between efficiency and code size (which can be a big deal).

If you are implementing a cryptographic algorithm from a specification, you can make some design decisions about how the algorithm is implemented, but remember to make sure that none of the optimizations cuts any corners that may lead to a breach of the integrity of the security provided by the algorithm. If you are using a prewritten library or source code, you may be limited to the options for optimization provided by the original author. When we look at building a secure application using a Microchip PIC processor, we actually utilize an open-source implementation of AES ported to the PIC. The implementation actually allows us to choose between a compact and slow variant and a larger, faster version, controlled with macro pre-processing. On the PIC, code space is plentiful, but RAM is limited. We will look at how we make the implementation of AES work on the limited-resource PIC in Chapter 11.

We can optimize certain algorithms to get a little more speed out, but unfortunately, there is only so much that we can optimize. In many cases, pure software optimization may be enough, but some applications require better performance than can be provided. One obvious choice is to move the critical operations into hardware, but that may not be an option. In the next section we will look at some performance differences between different algorithms and how algorithm choice may give your application the performance it needs.

Choosing Cryptographic Algorithms

In the last section we discussed the potential for optimizing algorithms, which can be done, but sometimes may not result in the type of performance required. As was mentioned (and covered next in Chapter 9), you can always move the critical cryptographic operations into hardware, but this may not always be possible, especially considering the additional expense and the fact the design is locked into a particular scheme for encryption. In fact, this was exactly the problem with the Wired-Equivalent Privacy (WEP) protocol originally implemented for the 802.11 wireless network protocols. Many 802.11 applications utilized specialized hardware to speed up WEP, but once WEP was discovered to have some serious security flaws, it was impractical to update all that hardware. The solution (WPA) was implemented to utilize the existing WEP hardware, but this limited the capabilities of the new protocol and took a rather significant effort to implement.

Fortunately, there is a way to get the performance you need for your application without some of the drawbacks of a hardware-based solution. The answer is to design algorithms with a natural performance advantage into your application. Obviously, this will not always work since you may have external requirements that dictate the use of particular algorithms. However, if you do have the choice, you can sometimes trade security for performance. There are enough algorithms available that you actually can choose from a relatively wide array of performance characteristics. Unfortunately, this can also be a tradeoff, since additional security often translates to more repetitions or more complex math. The RC4 cipher algorithm, for example, is extremely fast and simple to implement. Unfortunately, RC4 is usually considered less secure than more rigorous algorithms like AES. It must be noted, however, that there are no known practical attacks on RC4 as long as it is used properly (remember the caveats to using a stream cipher). The WEP protocol used RC4, and this is often cited as a reason why the protocol failed, but the problem was in how WEP generated keys in combination with RC4.[2] If you are careful, you can still utilize RC4 to provide both high performance and a relatively acceptable level of security. In fact, SSL implementations use RC4 more commonly than most other symmetric algorithms (see Chapter 4)—looking at how SSL uses RC4 and generates keys would be a good starting point to implementing an appropriate RC4 solution.

The other common symmetric algorithms include AES and DES. DES is essentially obsolete, replaced by AES, but it may still be required for communication with some legacy applications. DES is a dinosaur and can be quite slow when implemented in software, since its original design was optimized for implementing in hardware in the 1970s. Its original replacement, 3DES ("triple-DES") is simply a method of running the algorithm over the same data three separate times, each with a different piece of the key. The key was essentially 3 DES keys strung together, and each operation used a third of the larger key. Since 3DES was the same algorithm run three times, it can be expected to be three times as slow as plain DES, making its performance characteristics even worse. 3DES, like DES, is essentially obsolete, though its significantly larger key size provides some additional security.

Compared to DES and 3DES, AES is a much more modern algorithm, and it was even designed with performance on limited-resource platforms in mind (part of the AES competition rules stated that the algorithm should be easy to implement with reasonable performance on all machines from tiny 8-bitters to supercomputers). At this point in time, choosing AES is generally a good idea for any application, since it is a standard, commonly used algorithm that provides some of the best security currently available. AES is also

---

[2] From an RSA Laboratories technical note; *RSA Security Response to Weaknesses in Key Scheduling Algorithm of RC4* (www.rsa.com)

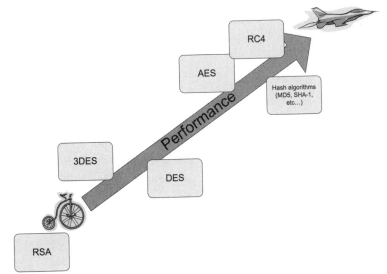

Figure 8.1: Relative Performance of Common Cryptographic Algorithms

scalable to different key sizes and the number of rounds can be increased to increase the relative security provided. The reason AES can be implemented efficiently is due to the fact that the algorithm is essentially modular, and can be implemented in different ways that take advantage of the underlying hardware. That being said, AES is still significantly slower than RC4, but for compatibility and security, it really cannot be beat for applications requiring symmetric key encryption.

AES, RC4, and other symmetric-key algorithms are fine algorithms as long as you can arrange to share a key either before the application is deployed or somehow distribute keys afterwards. For many applications, especially embedded applications, distributing a key after deployment can be extremely difficult or simply infeasible. Presharing a key is not often an ideal situation either, since the keys are either difficult to manage (which device has which key?) or if the keys are all the same, there is a serious security hazard (if one device is compromised, they all are). With these inherent difficulties in key management for symmetric cryptography, it is sometimes necessary to employ an asymmetric (public-key) algorithm in order to have the ability to handle key management without having to pre-share symmetric keys. In some cases, it may even be useful to have the public-key algorithm handle all data transfers. An example of the use of public-key cryptography for all data transfers would be a system where the amount of data to be sent was very small (tens of bytes) and sent very infrequently, such as an irrigation controller that usually runs in an automated fashion, but occasionally needs to be overridden. If code space is extremely limited (such as on a PIC), and speed is not a factor, implementing a single cryptographic algorithm may be desirable, and an algorithm like RSA can be implemented in a small amount of code. Generally speaking, however, public-key algorithms are so slow and cumbersome in execution that they are rarely (if ever) used as the sole encryption method. More often, they are paired with more efficient symmetric-key algorithms in protocols like SSL or SSH.

Public-key algorithms are usually rolled up into a protocol like SSL, but that does not mean your application has to implement a complete security protocol. However, if you need public-key cryptography for your application and you aren't sure about needing a full protocol, you probably should opt for the full protocol, since it will be more generally useful and may save headaches later (especially for compatibility with other applications). If you absolutely must have public-key algorithm support and cannot implement a full protocol like SSL, then it should be fairly obvious how to proceed. The public-key algorithm is used for distributing the key and the symmetric-key algorithm handles the bulk of the communications encryption.

The astute reader may have noticed that the public-key algorithm does not completely remove the requirement of pre-sharing keys, since each device may require a private key. It would be possible to make each device a "client," which would allow a central machine to send out public-keys to each device (the central system has the associated private key), but this may not be feasible for many applications. However, even if each device requires a private key, the only problem is at deployment when each device is assigned its own key. Once deployed, the device can simply provide its public-key to the client at any time, thus obviating the need for any key management. For more information on how public-key algorithms are used with symmetric-key algorithms in practice, a good place to look is Chapter 4 on SSL—the SSL protocol is a good example of the pairing of algorithms to provide the benefits of each type.

We have so far discussed the basics of how public-key algorithms are used, but the real issue is their performance. The problem is that the nature of public-key algorithms requires far more computationally intense mathematical operations than their symmetric cousins. The reason for this is the fact that the security they provide is based upon the difficulty of factoring large numbers. Modern computers are good at doing factoring by using brute-force methods on small numbers, so the numbers employed by public-key algorithms must be large enough to ensure that there will be little possibility of someone being able to factor them quickly using a computer. The result of this requirement is a much larger key size than symmetric-key algorithms for an equivalent amount of security. As an example, 512-bit RSA is considered to be about the same level of security as a typical 64-bit symmetric operation. This order-of-magnitude difference in key size translates to a similar order-of-magnitude difference in performance. The reason for this is that part of the RSA key is actually used as an exponent to derive an even larger number. Now, it should be fairly obvious that if you were to apply a large exponent to another large number, the size of the number would grow extremely rapidly. However, RSA works on the principles of modular arithmetic, so the result is truncated by the modulus (part of the public-key) after each multiplication. Even though we can store the truncated results, all this multiplication takes an extremely long time and even being able to do millions of operations a second, the result is still a significant slowdown. Without hardware assistance, a 512-bit RSA operation can take 30 seconds or more on an embedded microcontroller running at 50 MHz.

The RSA operation for encryption and decryption is relatively simple as compared to other cryptographic algorithms, but the magic of RSA lies in the keys. Key generation in RSA is less simple than the encryption operations, but it reveals the security of the algorithm. Each RSA key is derived from a couple of large prime numbers, usually denoted as p and q. The public modulus n is the product of p and q, and this is where the security comes in[3]—it is generally considered to be impossible to determine the prime factors of very large numbers with any method other than brute-force. Given a sufficiently large number, even a modern computer cannot calculate the prime factors of that number in a reasonable amount of time, which is usually defined in terms of years of computing power. A secure algorithm typically implies thousands or millions of years (or more) of computing power is required for a brute force attack. The trick to RSA is that the private key (d) is derived from p and q such that exponentiation with modulus n will result in the retrieval of the plaintext message from the encrypted message (which was calculated by applying the public exponent e to the original message). All of the mathematics here basically leads to the result that RSA is slow, and something needs to be done if it is going to be utilized on a lower-performance system. As we mentioned, hardware assistance is really the only way to really speed up the algorithm, but there is a method that utilizes a property of the modular math used by RSA—it is based on an ancient concept called the Chinese Remainder Theorem, or CRT. CRT basically divides the RSA operation up using the prime factors p and q used to derive the public and private keys. Instead of calculating a full private exponent from p and q, the private key is divided amongst several CRT factors that allow the algorithm to be divided up into smaller operations. This doesn't translate into much performance gain unless it is implemented on a parallel processor system, but any gain can be useful for a relatively slow, inexpensive embedded CPU.

Using CRT to speed up RSA is pretty much the only known software method for optimizing RSA—it is simply a CPU-cycle-eating algorithm. For this reason, hardware optimization is really the only choice. This was recognized early on, and in the early 1980s there was even work to design a chip dedicated to RSA operations. Today's PCs are now fast

---

[3] This description of the RSA algorithm is derived from "PKCS #1 v2.1: RSA Cryptography Standard (June 14, 2002)," from RSA Security Inc. Public-Key Cryptography Standards (PKCS), www.rsa.com.

enough to run RSA entirely in software at the same level of performance as the original hardware-based solutions. However, given that a large number of modern embedded processors possess similar resources and performance characteristics to computers in the 1980s, we still have to contend with a performance-stealing algorithm. RSA is the primary reason there are few implementations of SSL for inexpensive embedded platforms. The few that do support SSL stacks typically utilize some form of hardware assistance. In Chapter 9, we will look at some of the options for hardware assistance that are available for inexpensive embedded systems and how they can be utilized to improve the performance of cryptographic algorithms.

Tailoring Security for Your Application

Before we take a look at hardware-based security optimization, we will spend a little time talking about tailoring security to your application. Generally speaking, SSL/TLS is going to be your best bet for a general-purpose protocol, since it provides a combination of authentication, privacy, and integrity checking. For some applications, however, the security requirements do not include all three security features provided by SSL. For example, you may only need integrity checking, but not privacy (you don't care that attackers can read what is going across the network, but you don't want them to tamper with it either). In this case, hashing alone will probably be sufficient to provide the security you need. In other cases, authentication may be all that is needed, or some combination of privacy and integrity checking. However, if you think about the security provided by SSL, you will see that the three security properties it provides are intertwined, without some privacy in the form of encrypted messages, authentication can be difficult, and without authentication, integrity checking really only helps to prevent corruption of data, not directed attacks. For this reason, you will see that the properties embodied by the SSL and TLS protocols are inherent in any good secure application. When we look at the PIC in Chapter 11, you will see that even though the primary goal of the encryption used by the application is to prevent an attacker from changing the system state, the protocol implemented therein provides both privacy and integrity checking in addition to the authentication mentioned.

When implementing your own security solution around a cryptographic algorithm, instead of using a publicly available protocol, you should remember the lessons provided by SSL and look to implement privacy, authentication, and integrity checking as a sound engineering practice, even if it does not seem that they are all needed.

One last thing to consider when tailoring security to your particular application is the network in which the device will be deployed. If your application needs to communicate with existing systems, then you will need to know what protocols are provided on those systems and what you will need to implement. If your application is completely under your control, then implementing your own protocol may save a lot of space. Be careful, though, since you will be walking down a dangerous path. If you need a high level of security, rolling your own protocol is almost never a good idea.

Wrap-Up

Throughout this book we have covered various protocols, algorithms, and strategies for securing applications that utilize network communications. In this chapter, we have looked at how we can optimize different algorithms safely for systems with limited resources. We are getting closer to the case-studies, where we will look at some applications in action, but first we will take a look at a couple of things that may be useful. In Chapter 9, we will look at some hardware-based solutions and hardware assistance features that may help to take some of the burden of processing off of an otherwise limited application. In Chapter 10, we will look at some miscellaneous security issues that are applicable to embedded systems design such as physical security, programming gotchas, the role of programming languages in security, and recognizing and dealing with attacks. These next two chapters are not necessary for understanding the case studies, so feel free to skip ahead to Chapter 11 and take a look at the PIC application.

CHAPTER 9

Hardware-Based Security

High Performance in Silicon

In Chapter 8 we looked at optimizing software implementations of cryptographic algorithms. In this chapter we will look at hardware alternatives that allow embedded developers to get the most out of their systems.

First, we will look at the topic of hardware assistance. With the advent of the Internet, and the increasing need for security, embedded CPU manufacturers have started adding special instructions to their chips in order to help speed up software implementations of cryptographic algorithms. We will look at some examples of instructions that are useful for common algorithms, as well as some real-world examples of embedded processors that provide this added functionality. We will look at the features of these processors, and look at a strategy for determining which features are important for what types of applications.

Hardware assist provides the software engineer with hardware resources to handle some of the more expensive cryptographic tasks, but how are these instructions and resources used? We will look at some sample code of software implementations that utilize these resources on real-world processors, and discuss how an engineer can go about developing his or her own optimized implementations. We will look at assembly programming and compiler-supported optimizations, as well as out-of-the box solutions.

We also need to look at complete hardware security solutions. We will look at complete IC packages such as the Atmel Trusted Computing Platform, and components that can be added to an FPGA device or processor to provide complete security support in hardware. We will also look at some other hardware features, such as write-once key registers and physical protection mechanisms.

Finally, we will look at on-chip versus off-chip solutions. There are advantages and disadvantages to each, such as on-chip solutions preventing bus monitoring, but using up valuable CPU silicon.

Hardware Assistance (Specialized Instructions/Mechanisms)

The concept of hardware assistance for math-intensive operations is as old as the modern computer. Today's processors now include instructions that speed up floating point calculations, something that used to be included in expensive co-processors. We still implement operations in hardware that could be done in software but benefit from the enormous performance gain. We will cover complete hardware solutions later in this chapter, but first we are going to look at partial solutions that focus on more general types of operations that can be implemented directly in hardware. These operations do not make up the entire operation, but rather speed up some common operation that would take much longer in software.

One of the primary examples of hardware assistance for cryptography in a CPU is the inclusion of an instruction or instructions that perform repeated operations in hardware, rather than relying on some software mechanism. As an example, the RSA public-key encryption algorithm consists of raising a large number to a large power with a modulus. Thus, the operation is essentially an enormous number of multiplications of large numbers. The final result and all intermediate results are limited to the size of the modulus, so we don't have to worry about memory, but the number of multiplications (in RSA this would be determined either by the public exponent or the private key) is bounded by the number that can be represented by the key size—for even 512-bit RSA, this can be millions of operations (primarily with the private key). As a result, the RSA operation can end up being very slow. Compounding the issue is the fact that the numbers we are dealing with in RSA operations are not supported by the native instruction set of general-purpose processors—the RSA numbers are many bytes long (512 bits = a 64 *byte* integer!). Since the largest numbers supported by today's processors max out at 64 or 128 *bits*, any RSA software implementation must break up the multiplication operations on these long numbers into a number of smaller operations.

To speed up RSA, we could implement the entire operation completely in hardware (which is an option we will discuss in the next section), but this uses valuable silicon, so a less-intensive operation may be desirable. One thing that is extremely common in RSA is the multiplication of two large numbers (remember, 64 *bytes*—minimum!). Splitting that operation up into pieces that can fit into registers will take a lot of instructions (or a lot of repetition of a few instructions, anyway), and the instruction fetch alone will start to impact the end performance. At speeds in the tens or hundreds of megahertz (at which many embedded processors run at today), the combined effect can lead to a delay of several seconds—often too long, especially if you are waiting for a web page to load. The key is to provide instructions that can do the repetitive work for you.

For RSA, one of the most obvious bottlenecks is the multiplication operation. If implemented in software it requires that the numbers be loaded into registers a few bytes at a time (depending on the processor register size and instruction set) and multiplied appropriately, a piece at a time. Then, any carry value from the multiplication needs to be saved temporarily while the result is stored and the next pieces of the numbers are loaded. What if the processor had an instruction that did exactly that and repeated as many times as you wanted? This is obviously a CISC-style instruction, since it would take a variable number of CPU clock cycles to execute, but it could be implemented as a peripheral device if the CPU was RISC-based (we discuss this option later). In any case condensing the load, store, multiplication, and addition operations into a single load-once-and-go instruction can result in a speed up of 10 times or more! The reason for this is that you have eliminated the additional memory accesses that would be required to reload each of the instructions that would otherwise be required. An additional bonus is that the intermediate results are handled internally. Your program would no longer have to keep track of the temporary information between operations.

The real reason to look at hardware assistance as opposed to full-blown hardware implementation of your algorithm is to avoid using valuable silicon, but it also means that you can find operations that are useful for purposes other than for what they are primarily intended. This also means that if for some reason your encryption algorithm needs to change later, there is a better chance that the general-purpose hardware will be reusable, whereas if you had implemented that algorithm entirely in hardware, you would be stuck with it.

Note that the hardware assistance does not necessarily need to be implanted in the same silicon as the CPU, it would be reasonable and nearly as fast (or faster in some cases) to implement the hardware-assist logic as a peripheral. For this to work, the hardware would be implemented the same as any another peripheral in the system, with some interface accessible to the CPU core (the interface could be some type of control registers or direct memory mapping of the peripheral). In this case, the expensive (in clock cycles) operation would be offloaded from the CPU itself. The CPU would copy any appropriate information to the peripheral and then initiate the operation. The peripheral could then fire an interrupt whenever it completed the operation, thus freeing the processor to do other work. A simpler and perhaps less dangerous implementation would have the CPU simply wait for the result of the operation. In any case, the peripheral option would keep the instruction set smaller or allow for a RISC-style instruction set while still providing the performance gain of hardware assistance.

For a real-world example of hardware assistance in action, we will look at the Rabbit 4000 microprocessor.[1] The Rabbit, originally based on the venerable Z80 instruction set, has

---

[1] At the time of this writing, the author is employed by Digi International, owner of the Rabbit brand of embedded processors. We also cover the Rabbit 4000 later in a case study (see Chapter 12 for more information).

evolved into a more versatile processor, and includes many options not found in its ancestors. One of the notable features is the inclusion of some hardware-assistance instructions for cryptographic operations. One of these is the so-called "sbox" instruction, which provides access to an sbox, a lookup table used by AES implementations. The lookup table would otherwise have to be implemented as an array, slowing the AES algorithm considerably.

Other notable cryptographic hardware-assist features of the Rabbit 4000 are the unsigned-multiply-and-add (or subtract) instructions, known as UMA (and UMS). These instructions enable a greater than 10 times speedup in RSA operations over a full-software implementation using the other Rabbit instructions. The UMA instruction, specifically, is designed to make the operation of adding two arbitrarily large unsigned integers very fast. Basically a block-copy instruction with some additional logic, UMA can be used to make cryptographic operations much faster, particularly those operations used by the RSA algorithm.

The speedup from the UMA instruction is significant enough to allow the Rabbit, an 8-bitter running at less than 60 MHz, to perform an RSA decryption quickly enough for use in HTTPS (HTTP over SSL) for web-based interfaces (about 2 seconds for 512-bit RSA).

Hardware assistance is great for providing a level of security to a class of hardware not normally capable of the higher-level processing required by modern cryptography, but if you want to be serious about security, you need to spend a lot more money on a more powerful processor or a little bit more on a specific hardware solution. In the next section we will look at full-blown hardware solutions for security, where entire algorithms are implemented entirely (or mostly) in hardware, either as part of a larger general-purpose processor or as a standalone external unit that needs to be interfaced to a CPU.

Hardware-Based Solutions

Modern PCs are so fast that software solutions are usually acceptable (less than a half-second for an RSA operation is plenty fast for most applications). In the world of embedded devices with limited resources, implementing a full SSL stack in software may not be easy. Hardware assistance can get you part of the way to a secure system, but there are still limitations. For embedded applications where you need high speed and high security, you have two options: Go for a bigger, faster CPU or implement the security in hardware. We could talk about security on a multi-GHz embedded Pentium, but where is the fun in that? Security for fast processors is a subject for more general security books, and we are interested in having security without breaking the bank so we will focus on security in hardware.

One very feasible option for a hardware-based security solution would be to implement an SSL stack or cryptographic algorithms on an FPGA. This has the benefits of being customizable, and the cost of an FPGA can be less than a more powerful CPU (obviously this depends on what you are doing). FPGA programming is beyond the scope of this book, but it is an interesting idea. Most likely, you would want to implement the cryptographic algorithms for the FPGA and leave the full SSL stack in software. Using an FPGA should result in a fair gain in performance, but it will not be as fast as something forged in silicon. Fortunately for us, there are a few vendors that offer security-on-a-chip solutions.

The idea of putting security algorithms in silicon is not a new one. When RSA was first developed, there was an effort to design a chip dedicated to the algorithm. While the idea is less relevant for mainstream computing now that PCs have processing power that was unthinkable in the 1980s, the concept of putting cryptography into hardware still makes sense for embedded applications. For the types of applications we have been discussing, it is almost a necessity in certain cases.

A real-world example of a security-on-a-chip solution is Atmel's Trusted Platform Module, which provides a full RSA accelerator (2048-bit RSA in 0.5 seconds) along with a random number generator and nonvolatile memory for key storage. At only a few dollars each, this is a serious alternative to using a faster processor. With the additional key storage and random number generator, it would be a useful addition even to an application that used something more powerful.

Other vendors offer similar technologies, but they are not extremely common due to a lack of need. When most processors (even the embedded ones) can do an RSA operation in less than a couple seconds, there really isn't much need for hardware acceleration except under the most extreme circumstances. As we have mentioned throughout this text, security is truly dependent on the application. Adding another chip to your design may not be practical, and it comes with its own problems. One of the major issues is actually physical security, which we will talk a little bit more about in the next chapter. The issue is that you are placing a chip somewhere else on a circuit board, and there needs to be an interface to that chip. If your device is going to be deployed out in a field somewhere, what is there to stop someone with a logic analyzer from tapping into that interface and grabbing information? If the keys are stored in memory and transferred to the processor, the data bus is also a target.

With the relative scarcity of security hardware solutions that fit the bill for limited-resource embedded applications, you are probably better off just living with the performance hit of a software implementation or just pay more for a faster CPU. Hopefully, the tricks in this book will be enough to get you going on a solution that works for you without having to resort to specialized hardware (it just kind of violates the spirit of implementing security on a shoestring). Another issue is that security is a fast-changing field, and if you are locked into a particular hardware-based solution, it gives you very little room to adapt your application.

Wrap-Up: Security and Hardware

This chapter has been a very brief introduction to the concept of security in hardware. The information was included to give the reader a view to what is out there, but this book is primarily focused on security in software. Hardware-based security definitely has its place, especially with low-power wireless devices running ZigBee, but for most applications that kind of performance just is not needed. If you need fantastic performance from your security protocols and algorithms, you need better hardware, plain and simple. Cryptography is expensive and the overhead needed for many protocols can be immense. Hopefully, however, this book has given you some tools that will allow you to make your applications a little more secure. Hey, a little security is better than none, even if all it does is keep out casual onlookers.

In the next chapter, we will look at various security issues pertaining to the embedded systems industry, before continuing to the case studies in Chapter 11.

CHAPTER 10

Conclusion—Miscellaneous Security Issues and the Future of Embedded Applications Security

This book has been about securing embedded applications where resources are limited and cost is an issue. Before we delve into a couple of real applications in Chapters 11 and 12, we are going to cover a variety of topics that do not fit into any of the previous chapters, and may only be peripherally related to embedded applications, but should at least be mentioned. This book is targeted at relative novices in computer security, so this chapter is intended to be a springboard into other research. In this chapter, we will briefly cover a few of the more esoteric aspects of security, including the impact of programming language choice on security, dealing with and recognizing attacks, and the future of security, including quantum cryptography and legal issues.

Programming Languages and Security

Choosing a hardware platform and designing security features are only part of the work in developing a secure application. Development tools and languages factor into the security. Some languages, such as Java, provide some built-in safeguards that help with overall security, such as strict typing. The features of different languages and compilers are something to look at when choosing the software for your system.

The predominant language for most embedded applications is some variant of the C language. Unfortunately, C suffers from some inherent problems that detract from overall security, notably weak typing and problems with buffer overflow. There are some techniques to avoid common pitfalls, however, and we have looked at some of those throughout this book. C is such a popular language for embedded developers because of its close ties to the underlying hardware, but this can detract from the security of an application because it lets programmers do things they should not. The C++ language, being a descendant of C, also suffers from similar shortfalls. C++, however, has better typing (though not perfect) and a good object-orient design approach can help with the buffer overflow and pointer issues found in C. In any case, using C or C++ makes security a difficult job, but there often isn't a choice. Applying good engineering principles to your design is pretty much the only way to avoid the security pitfalls in C and C-like languages. Fortunately, however, almost every security protocol or algorithm that you may want to use will inevitably have a reference implementation in C.

An alternative to C that has been gaining in popularity is the Java language. Java is good from a security perspective because it enforces fairly strict typing and virtually eliminates some of the buffer overflow problems seen in C. Java exception handling is also quite nice for handling errors that represent possible attacks. The problem, though, is that unless you have a CPU equipped to run Java Bytecode directly, you have to rely on a Java Virtual Machine (JVM), which is most likely written in C. If you can verify that your JVM is secure, then it is easier to verify your application. Another problem with using a JVM, besides the security issues, is performance. Since Java is a partially interpreted language, it suffers tremendous performance penalties for certain applications. For this reason, Java is typically only available on higher-performance platforms.

Other than C (and various flavors of C) and Java, the language that may be most familiar to the reader is BASIC, available for some hobbyist-targeted hardware, such as the Parallax BASIC Stamp. BASIC does not have the same backing as C and Java, and it hides a lot of functionality (such as typing) from the software engineer. It should be possible to implement various cryptographic algorithms and even full security protocols in BASIC, but it is likely that there isn't the wealth of reference code available that can be found for languages like C and Java.

You should also look at the development tools themselves. If your compiler generates incorrect code, you may introduce a security hole without knowing it. If you are using inexpensive development tools, there is a greater possibility of incorrect code generation, but compilers are complex applications and even world-class tools can still suffer from problems. Inspect generated code in critical sections of your applications and be sure to keep up with the latest bug reports for the tools you use.

There are numerous languages that provide software support for different hardware platforms, but other than the languages mentioned here, they are likely aimed at niche markets or are specific to a particular vendor. When using a proprietary language to develop a secure embedded application, be sure that you understand what the language and the development tools are doing. The better you understand the languages and tools that you use, the less likely you are to make mistakes that will lead to security breaches later.

Dealing with Attacks

Any system connected to a network (the Internet or proprietary) *will* be subject to attacks—both intentional (i.e., malicious hackers) and inadvertently (i.e., heavy network traffic leading to Denial of Service—we have classified DOS as an attack throughout this book). Here we will briefly touch on what an attack means, and what we can do to identify attacks and deal with the aftermath of a successful attack.

The first line of defense is a good security policy and verification or assurance of the security of the implementation of that policy. If you can be relatively certain that you application and policy are sound, then you have likely prevented the lion's share of future attacks. We have covered the engineering principles and security mechanisms throughout the book, and implementing them is the first step to a secure system. However, no matter how well you implement your application there will likely be a problem at some point that will allow an attacker to gain an advantage. We have really only covered the prevention of attacks, but how do we recognize an attack? How do we deal with an attack that is happening? Both these questions are extremely difficult to answer and are far beyond the scope of this book, but we will talk a little about what can be done.

Recognizing that an attack has occurred or is occurring can be extremely difficult. Eavesdropping or man-in-the-middle attacks are impossible to detect, since they happen outside the system that is under your control. However, protocols like SSL will let you know when a possible attack is occurring (in the case of SSL, the attack notification is called an *alert*). Your application must treat every possible attack as the real deal, or you will eventually be attacked without knowing it (this is sort of a variant of the boy who cried wolf). Attacks that exploit your own system are a little easier to recognize, especially after the fact. If your application is behaving erratically, there is the possibility it was attacked, but remember that an attack and a defect can have the same symptoms, so don't jump to any conclusions. How you deal with an attack that has already occurred depends largely on what your application does. When designing your system, pay close attention to the actions you don't want to happen. Treat each of those actions as the result of an attack, and working backwards, identify as many paths to those end results as you can. This exercise will help greatly in developing a more secure application.

Handling a current attack is a different matter. Even if you have high confidence that your application is safe, there may still (and likely do) exist vulnerabilities that can be exploited. For this reason, a mechanism for intrusion detection may be desirable. As in the exercise mentioned above, you should identify any paths that lead to an action that is undesirable. If you cannot prevent an action from occurring, such as a Denial-of-Service condition, you may want to consider adding some logic to detect the early stages of an attack and react accordingly. However, designing an intrusion-detection system (or IDS) is trickier than it sounds. IDS development is a big business and a difficult problem to solve, even on machines with the horsepower to run complex programs that can identify various intrusions. An IDS can actually harm your application as well, if implemented improperly. As an example of this, imagine an application that stops accepting packets for a set period of time as a way to mitigate the effects of a Denial-of-Service attack. This may seem logical at first, but if the attacker knows the period of time used by the countermeasure logic, then it is even easier to successfully attack the system. If you want your application to be able to respond to attacks, be very careful in what you implement—in some cases, the solution may be worse than the problem.

Finally, there is the last resort. After all else has failed and an attack is successful, what can be done to recover? There are many different ways to handle the aftermath of an attack, and nearly all of them are application-specific. Basically, recovering from a successful attack will generally amount to damage control. We won't consider the business or human factors here, but there are some things you can do for your system. First and foremost, you will want to update other systems to prevent repeat attacks as soon as possible. Using redundancy (if one device is compromised, protect a redundant copy) can help to heal a broken system, and frequent backups help in recovering lost data or application state. Generally speaking, however, your application should be designed to prevent attacks, not recover from successful ones. Doing so would be the equivalent of preparing to deal with being mugged walking down the street, rather than taking measures to prevent being mugged in the first place.

The Future of Security

Security is a constantly evolving field. Truly the only way to provide any level of security for an application is a constant state of vigilance and attention to developments within the security community. With prevention being the best solution, knowledge is your best weapon. The Internet is a fantastic resource, since information on many attacks and security flaws is almost immediately accessible. Security watchdog organizations like CERT provide up-to-the-minute updates on the state of the Internet and what critical problems are currently known. Hopefully this book is a good starting place for you to learn about security, but it should not be the only place. Buy as many books as you can and read security-related blogs and forums. The only way to keep ahead of the hackers is by being prepared.

So what about the social and political aspects of security? Well, it turns out that computer security presents some very interesting challenges for society. Up until recently, most encryption algorithms above a certain key strength were considered munitions by the United States government and therefore all exports of source code were prohibited. Ironically, foreign governments (Israel, Russia, etc.) independently developed cryptography that was equal in strength to anything the US had. In the early 2000s, the US reduced the export limitations on encryption and applications using encryption, but sending a device with 128-bit AES to countries like Syria or Iran is likely still very illegal. If you live in a country other than the US, or you are in the US and send products to other countries, be very careful about the laws and regulations regarding encryption—it may be very easy to get yourself into trouble without even knowing it.

Another social issue with computer and electronic security that will not be resolved anytime soon is the problem of strong encryption falling into "the wrong hands." With AES reference implementations easily downloaded, anyone with a connection to the Internet can access strong encryption to protect their communications. This presents a paradoxical challenge—without security mechanisms, we cannot keep our own data and communications safe, but if we make it publicly available (which is a good idea as we have already mentioned), then our enemies can use it against us. The fact of the matter is that source code and computers are much easier to obtain than enriched uranium, so any focus on preventing the enemy from obtaining electronic security is inevitably futile. Like dealing with attacks, the best solution to this problem is prevention, but that we will leave to the diplomats.

Social issues aside, what then is the future of security technology? Well, part of the answer is new types of encryption algorithms that offer different properties. One relatively newer class of encryption are the so-called "elliptic curve" algorithms, which derive their security from the mathematic principles of (you guessed it) elliptic curves. These algorithms tend to be fairly fast and represent a different way of hiding information, but they have not received the same attention as the older methods. Time will tell if elliptic-curve cryptography will become widely adopted by the industry.

Elliptic-curve encryption is a new method for an old way of doing things, so the future of security may actually lie in a completely new way of looking at security. One of the most exciting (and potentially frightening) areas of interest for security researchers today is quantum computing. Utilizing quantum mechanics as a computation mechanism, quantum computers will theoretically make almost all existing forms of cryptography obsolete overnight. For now, a full quantum computer is purely theoretical,[1] but if implemented, it should be able to factor large numbers in a fraction of the time any conventional hardware could ever dream of achieving. As we mentioned several times throughout this book, the security of RSA is based upon the relative difficulty of factoring large numbers. Quantum cryptography would make factoring easy so no matter what we did, RSA would no longer be secure since keys of any reasonable size could be discovered with relatively little effort.

Quantum computing represents a relatively apocalyptic future for computer security, but fortunately, all is not lost. First, no one has been able to build the theoretical holy grail of quantum computers—no one is even very close. It turns out that the quantum properties that (theoretically) make the complicated math really easy are not that easy to control.

---

[1] Well, the NSA might have a quantum computer, but they aren't telling anyone about it!

Manipulating individual atoms is a hard problem, and it may even be impossible to build a full quantum computer (but remember, the designers of DES thought it would never be broken either!). Second, quantum computing itself may offer the answer we are looking for to protect our data in the event that quantum computers become commonplace. This off-shoot of quantum computing, called quantum cryptography, utilizes the same quantum properties as quantum computing to encrypt data. As a bonus, some forms of quantum cryptography are much easier to implement than a full quantum computer. Generally speaking, quantum cryptography has a useful property. Eavesdropping protection and tamper detection are inherent in the quantum properties being exploited. The current best-known method for quantum cryptography utilizes single photons traveling over a fiber-optic cable (some companies already sell devices they claim use this method). The quantum "spin" of the photons is used to encode data. The receiver is able to put together a string of photons to produce a message. However, by the strange properties of quantum physics, if an attacker were to "read" the spin of a single photon in the string, it immediately changes the properties of the photon, thus destroying further information and simultaneously alerting the receiver that the communications channel has been tampered with. Despite its useful properties and the promise of a new order in computer security, quantum cryptography is still a very new concept and technology that utilizes quantum mechanics is extremely expensive. It will be a long time (if ever) before we see quantum devices enter the main-stream electronics communications industry—especially in the embedded systems space.

The future of security for electronic systems is wide open. It is nearly impossible to predict what the next couple of decades will do for the field and what advances will change every-thing we now know. One thing is certain, however, and that is the fact that security will need to be an integral part of all electronic devices and applications unless human nature is changed and we all start being nice to one another.

Wrap-Up

So far, we have covered both theoretical and practical approaches to security on embedded devices. The focus has been an introduction to various aspects of security and the chal-lenges faced in implementing secure applications with limited resources. This varied cover-age hopefully has provided the reader with a general understanding of the field of embedded security and given the reader the tools to move on to more detailed sources. Hopefully the reader has also gained some insight into the challenges in developing practical, secure applications. In Chapters 11 and 12, we take all of the knowledge imparted in the first 10 chapters and apply it to a couple of real, working applications utilizing actual hardware. In Chapter 11, we implement a secure console-style networked interface to an Ethernet-enabled 8-bit PIC processor using a readily-available AES implementation. In Chapter 12, we look at a higher-end 8-bit processor, the Rabbit 4000, which conveniently is supplied with source code and libraries that make implementing a secure system easier. In both chapters, we go through a complete (though abbreviated) design and implementation cycle, documenting how the applications were designed from the requirements phase into design and finally implementation. Complete source code for the applications (not including support libraries provided with the development environments) can be found in Appendix A.

Finally, in the spirit of further knowledge and research, we provide a list of recommended reading to further your knowledge of computer security. None of these titles is specifically geared toward embedded applications, but the security principles found therein will apply to any application you may want to secure. Good luck implementing your own secure embedded applications and happy hunting—it's a jungle out there!

Applied Cryptography, by Bruce Schneier. *The* definitive reference on cryptography. No security junkie should be without it.

Practical Cryptography, by Niels Ferguson and Bruce Schneier. A more practical and approachable text than *Applied Cryptography* by the same author.

Security Engineering: A Guide to Building Dependable Distributed Systems, by Ross Anderson. A comprehensive coverage of security in general.

Cryptography and Network Security, by William Stallings. A thorough coverage of network security and cryptography.

Network Security Essentials (2nd Edition), by Stallings. Another text covering network security and networking protocols.

www.ietf.com—Internet Engineering Task Force website. All the RFCs referenced throughout the book are available here.

www.cert.org—Carnegie Mellon University's Internet security response team website. CERT keeps an eye on up-to-the-minute security issues pertaining to the Internet and networking technologies.

www.schneier.com—Bruce Schneier's website. Generally a good place to get information on anything security-related.

www.nsa.gov—the National Security Agency of the United States. Has some history and interesting tidbits about security in general.

PIC Case Study

Microchip PIC with Ethernet Controller

The Microchip PIC family of microcontrollers is one of the oldest and most recognizable embedded hardware offerings available. Developed in the mid-1970s, the PIC was originally a simple 8-bit microprocessor for improving I/O performance. In the late 1980s, the original PIC was updated to be a programmable microcontroller with the addition of EPROM for storing program code and data. Today, Microchip offers a wide variety of PIC processors, from the venerable 8-bit microcontrollers to modern, fast 16-bit controllers capable of far more than their ancestors. PIC processors have limited resources, instead providing just the right amount of processing power for varied applications. On the low end, some PIC microcontrollers have less than a kilobyte of memory and have only a few external pins. On the high end, the 16-bit PICs provide a lot more processing power. In the middle are the high-end 8-bit processors that are capable of quite a bit, including serving web pages and various other applications.

There are pre-written libraries available for the 16-bit PICs for various security protocols, including many of the standard encryption algorithms and even SSL. In this case study we are going to focus our attention on the lowlier 8-bit PIC, which does not have the same prepackaged software support for security as its 16-bit brethren. The hardware used for our PIC case study is the PICDEM HPC Explorer Board, along with the optional PICtail Ethernet Daughter Board for network support. The HPC explorer board features a PIC18F8722 microcontroller with 128 KB of onboard flash memory for program storage, 3936 bytes of RAM, and 1 KB of data EEPROM. The PIC18F8722 can run at speeds up to 40 MHz, making it the hardware with the least resources available of the platforms we look at in this book. However, despite the limited resources, we can still do quite a bit with this little processor. Fortunately, it already has a complete TCP/IP stack available for the Microchip C18 compiler (which is a basic ANSI-C compiler that is designed for the 8-bit PIC controllers). What it does not have is a security package—the Microchip encryption libraries are only available for the 16-bit PICs for the C30 compiler. As a result, if we want to secure a network connection using this hardware, we will need to figure something out ourselves. We are going to focus primarily on the security aspects of the application, but for

Figure 1: PICDEM HPC Explorer Board

more information on networking with the PIC, check out *Networking and Internetworking with Microcontrollers* by Fred Eady.

The situation of rolling our own security brings us to OpenSSL. OpenSSL is an open-source implementation of the SSL and TLS protocols, complete with implementations of all of the standard encryption algorithms commonly used with SSL. We could attempt to port OpenSSL to the PIC, but the microcontroller we are using just does not have the resources to support a full SSL stack (actually, it should be feasible to implement SSL for even the small amount of available resources on the PIC18F8722, but it would require some significant engineering effort with a lot of optimized assembly coding and some clever tricks far beyond the scope of this text). Instead of trying to squeeze the entire OpenSSL stack (which can take as much as 1 MB of code space or more when compiled) into the limited resources of the PIC, we will instead focus on a single encryption algorithm and look at what we can do to secure a network connection without SSL.

The algorithm we will focus on is AES. AES was designed to use a limited amount of resources, and the algorithm is scalable for different levels of processing power and memory available. It can also be written entirely in C, and the OpenSSL implementation is very accessible, with no assembly required. Optimizing the AES algorithm with hand-coded assembly is beyond the scope of this example, but it represents the best way to get some additional performance out of the application if you are brave or experienced enough to tackle the challenge. Assembly optimization is especially important on a processor such as our PIC, since it is not the fastest CPU on the block. In any case, we do not need AES to be blazing fast for our example application, since we are sending a few commands using a

Figure 2: PICtail Ethernet Daughterboard

command-prompt interface. The application, described in the next section, tolerates a slower response rate since we are only turning LEDs on and off, rather than controlling something that requires quick responses. If you were going to design an application that needed both quick response time and encryption over a network, you either need to reevaluate the hardware you are using, or redesign the application so that the response time is not dependent upon the network or encryption, possibly by making the controller smarter or offloading complex calculations to a more powerful system. In any case, our example works just fine the way it is, but there are a lot of things that can be done to improve its performance. The primary advantages that could be gained would come from writing the AES routines in PIC assembly—this would make the code less portable, but much faster. The OpenSSL AES implementation also has an option to unroll the inner loops of the AES algorithm, which uses more code space but might make the processing faster. However, AES is already a fast algorithm, and it may not be necessary to optimize it at all. The case study exhibited no noticeable delays when sending commands to blink the LEDs.

PIC Example Application—Secure LED Blinking

Our example application for the PIC18F8722 microcontroller (with Ethernet capability) is not the most creative in terms of its concept. We are going to blink some LEDs using a secure network connection. However, the extremely simple example will help us to develop a simple secure interface for our tiny controller. The PIC is fairly limited in its resources, so we need to keep things from getting too complicated. AES is a relatively good choice here, since it was designed with small embedded systems in mind (it can be written to be small and fast). We do not need the full complexity of something like the Secure Sockets Layer,

Figure 3: Microchip MPLAB ICD 2 Debugger and Programmer

since it would use up valuable resources, but we want some type of security for the commands we are sending to our hardware—think of each LED as a sprinkler in an automated watering system. It should be possible to implement SSL for the PIC, there is nothing preventing someone from doing it (despite what 8-bit microcontroller manufacturers sometimes want you to believe), but in general it isn't the most practical method for providing a secure interface to your application.

Our application, rather than rely on a complex protocol, will simply use AES, nothing more. (Well, that is not entirely true, since we will be developing a simple protocol of our own, loosely based upon ideas from SSL.) However, we need to be careful in how we use AES or we will be subject to certain attacks that would violate our security needs. In order to define what our applications security needs are, let's look at how we want our application to function, and especially at how we *do not* want it to function:

- Our application blinks LEDs on the development board on and off, remotely using a network connection (via the Ethernet daughterboard).

- We *do not* want an unauthorized person or system to be able to change the state of any LED.

There, that's pretty simple, right? We can just use AES to encrypt messages to tell the board to blink the LEDs, send them to the device, and no one can do anything unless they have the AES key, right? Wrong. We need to be careful when we are just using an encryption algorithm to send messages, since it will not prevent an attacker from being able to modify the system state against our wishes. One of the primary problems is the *replay attack*, where the attacker will simply observe the behavior of the device when messages are sent to it, and assuming the attacker has access to the network (usually a good assumption to make when trying to secure a system), he records each encrypted message and makes a note of what that message does. Later, when the attacker wants to change the system state to further his nefarious goals, he simply sends one of the prerecorded messages to the device and the system happily changes state based on the contents of the message. The replay attack will be a problem unless we do something to make every message unique and unpredictable.

To protect against replay attacks, our primary concern is to provide a method to differentiate messages so that resending an earlier message will not result in a state change. In order to do this, each message needs to be relatively unique, even if the commands being sent are exactly the same. For our sample application, we will use several methods to make replay attacks impossible and help to prevent someone from using known information to derive the key (known-plaintext attacks). The first mechanism we will employ is a simple counter, synchronized between the PIC system and the PC-based client program, and it will help to make each message unique. The second security measure is the introduction of random bits into the message. When combined with the third security mechanism, a simple checksum, the possibility of a replay attack is greatly minimized (and impossible if the random bits are truly random). As can be seen below, we are only utilizing a couple bytes in the AES block for the command protocol, so it will be easy to add in the random data in a few bytes.

We could do something more elaborate, such as providing a time stamp and hashing the entire message (including the time stamp) using SHA-1 or MD5, and then encrypt the whole thing. For our application, however, our simple protocol will suffice. We will allow the counter to be reset using a command that is part of our interface (and secured using AES, of course). This will allow the user to put the system into a hard-to-predict state, thereby lessening the chances that an attacker could predict the value of the counter and somehow use that information to some advantage.

Our simple protocol is going to utilize a command "packet" data structure that is exactly 16 bytes in length (see format below). Why 16 bytes? Well, this is the size of a single AES block when the key size is 128 bits (128 bits = 16 bytes). We can therefore utilize just the block cipher of AES and not have to implement any stream methods on top of it. The stream implementations for AES, such as Cipher Feedback (CFB) and Cipher Block Chaining (CBC), make AES function as if it were a stream (byte-at-a-time rather than blocks). We are only transmitting single blocks, so there is no real need for the stream capability. Always sending the same size block has an added bonus as well—since the length of every message is identical, there is no information that can be inferred from the length of the encrypted data.

Now that we have looked at what we need to do to secure the application, we can revisit the parameters of our design and add in a few new requirements:

- LEDs can be turned on and off via a command sent securely to the network port on the device.

- AES will be used to encrypt the messages.

- A counter known by both the client (PC) and server (device) will be used to differentiate messages to prevent replay attacks.

- To help prevent other attacks, the device will support updating the AES key using a command sent securely from the client.

- To further aid in preventing an attacker from being able to discern any useful information from the encrypted commands, the device will support a command that resets the counter. Using this, the client can periodically change the counter to help prevent predictable patterns in the ciphertext.

- A message from the PIC is sent back to the PC client program upon executing a command. The implementation of this message is to simply encrypt a 16-byte message and the client program prints it out, thus verifying the command was executed. This idea (purposefully) has a couple of issues that could compromise the security of our application. The reader is encouraged to spot the error or errors before we reveal the issue and provide a solution a little bit later.

- Additionally, several bytes worth of random data will be added into the command packet structure, and a checksum will be calculated for the final packet. The checksum is one byte in length, and will be calculated from the remaining 15 bytes, including the commands, the random data, and the counter.

- As a potential final mechanism to aid in preventing denial-of-service attacks (also handy for making sure the device is functioning properly), the device could send periodic messages to the client (once connected) that indicate it is still "alive" and connected. The device would use the same counter and encryption scheme as the commands to prevent replay attacks using the alert. The interval between the messages should be fixed so that no one could inadvertently trigger a flood of messages (interval too small) or prevent the messages from being useful (interval too large). This wasn't implemented for this case study, but it is a possible improvement to the design that would not be too difficult to add.

Now that we have all the requirements for our application, let's take a look at the command protocol, which consists primarily of the 16-byte command packet. The command packet format is as follows:

- Byte 1: The checksum, which is calculated by the PC client and the PIC on the plaintext command and used to verify the incoming information.

- Byte 2: The LED control byte, which is a bit-field representing the 7 LEDs used in the example (the 8th bit is ignored). By convention, a 1 will represent an "ON" state and 0 will represent an "OFF" state.

- Byte 3: The command byte, which is a bit-field representing the different commands. In the implementation, there are 3 commands: change the AES key, reset the counter, and a command that makes the PIC ignore 0-bits in the LED control byte, thus preserving the state of those LEDs (this can be used to prevent inadvertent state changes).

- Bytes 4–8: Reserved bytes, which are for future expansion of the protocol and will always be set to 0.

- Bytes 9–12: Random data bytes, populated by the PC client program.

- Bytes 13–16: The 32-bit counter value.

Graphically, the packet looks like Figure 4.

The application is pretty straightforward in terms of implementation. The basic idea is to have a loop on the PIC that constantly monitors a TCP port for incoming traffic. The incoming packets will contain the encrypted command packet indicating the commands that are to be executed. Once the PIC program executes the command, it goes back to its monitoring state. One feature to consider is a response to the user (using the PC client program) that the command was properly executed—otherwise, the user may not know if the command was executed at all. The example described here has immediate feedback, assuming the PIC hardware is sitting on your desk—you can just look at the LEDs to be sure they turned on. For real-world applications, feedback may not be immediate, so a response will be necessary. However, in the event of an error or malformed packet, the PIC program should do nothing. If it sent a response to every incorrect packet, then an attacker could use that to set up a denial-of-service attack by flooding the device with packets containing random data. The lack of a response within a certain timeframe should instead trigger the client program to signal an alert to the user. The "completed command" response

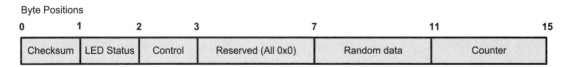

Figure 4: PIC Example Packet Format

mechanism provides a mechanism to defend against truncation attacks (or disconnected hardware), since the user will have nearly immediate feedback from the target device. Lack of response indicates that something is awry.

The described method of sending responses to commands should suffice for our example, but sometimes it will not be enough—it depends on what the application is and what it needs to do. One thing to note here is that the responses are sent by the PIC once the commands are executed. If this was a real-time system and you did not want an attacker to know what you were doing, the time interval between the PIC receiving the command and then sending the response could be used to gain information about the actions being performed. An example of this would be if the controller needed to do some mechanical task, such as reposition a missile turret. This obviously would take some time, perhaps several seconds, whereas another command—say, firing a missile—might complete within a few dozen milliseconds. The solution to this problem is to send all responses back to the client at a fixed amount time after the command is received. The interval here would necessarily be longer than the time to complete the command that takes the longest time to execute. Another possibility would be to send the responses at random intervals all at least as long as the time to complete the longest command. Both these methods would make it much more difficult for someone monitoring the network to determine what was happening. If this all seems a little excessive, it's not. This exact type of problem has led to some attacks on cryptographic algorithms such as RSA. The reason for this is that the RSA operation is essentially a huge number of multiplications strung together (an exponentiation to a very large power). By observing the time it takes a target to complete the RSA operation, an attacker would be able to narrow the range of possibilities for the key, without having to know anything else. The fixed-time response mechanism would be fairly straightforward to implement, simply start the clock when a command is received, and when the proper time period has elapsed send the response.

The response mechanism seems simple enough, even with the added security of the fixed-duration response mechanism, but we indicated above that there was an issue with the command response from the PIC to the PC client program. Were you able to figure out what the issues are? Well, the primary problem is that we are always sending a static message—the encrypted message will always be the same as long as the key is the same. This may sound innocuous enough, since the response is not really doing anything other than verifying that the command was executed. Also, you can always change the AES key occasionally, right? Well, there are two issues with the mechanism as proposed. First, it suffers from a known-plaintext attack vulnerability. A known-plaintext attack utilizes some known information (in this case, the static message. Keeping it secret is not sufficient to make the mechanism secure) to derive the key. It is not likely someone would go to the trouble of attacking a blinking LED application (or even a lawn sprinkler system), but we went through so much trouble to make the command structure secure that it is very obviously the weak link. A second, probably more severe problem (for this application, anyway) is that the mechanism suffers from a man-in-the-middle vulnerability. An attacker could intercept the response packets during normal transmission, save them, and then hijack the network to prevent traffic from reaching the PIC. The saved response packets are indistinguishable from one another, since they each contain the same message encrypted with the same key. Assuming that the key had not changed, the attacker could send a response to every hijacked packet indicating (falsely) to the PC client that the commands were being executed successfully when in reality, they were not being executed at all. This is essentially a truncation attack since not all the data is making it to the target—and this is exactly the type of attack we were trying to prevent!

The answer to our man-in-the-middle problem is to employ the same mechanism we developed for the command protocol to protect the response messages. Specifically, it would be good to create a "response packet" that contains the message (it could include the device state as a bonus feature), the global counter, and some random bits. Add in a checksum, and we have the same basic structure as the command packet. To make the implementation easier, we can simply reuse the command packet on the PIC. Simply update the counter, clear the command byte, update the LED byte with the current LED state, and recalculate the checksum. We can reuse the random data from the PC (which is likely better random data than we can get on the PIC anyway). The response packet then contains the LED state as part of the response, and we avoid the man-in-the-middle attack, since each response will be unique. If you are really paranoid, you could generate new random data on the PIC, and use a separate key to send the responses (this is essentially what SSL does), but this is probably overkill for our example. Just remember, though, it is better to be safe than sorry. If you adapt this example for a real application, you should probably implement all the recommended enhancements as long as you have the resources to do so. In any case, let's stop talking about what we plan to do and get to the implementation itself.

Implementation

We now know what we need to do to build our application, but where do we start? Well, the natural place to start, unless you are intimately familiar with the hardware and its development tools, is to start with a sample program that is similar in function or structure to the application you are looking to build. In our case, the TCP/IP stack for the PIC comes with a sample application that sets up a web server with an AJAX[1] interface to control LEDs and monitor the device status. The web interface is nice and spiffy, but we don't have a lot of resources on our PIC, so we will strip out most of the web server to free up

---

[1] AJAX is a Javascript-based framework used to provide low-overhead interactive content for web pages. One of the most recognizable applications of AJAX is in the Google Maps service.

resources for the encryption and the rest of our application. We could use the web server, but making it secure would be difficult, since the default security option for HTTP is SSL, and as we have already mentioned, this particular PIC is not really up to the task for full-blown SSL. Stripping out the web server is thankfully very easy, as we will see next. The example was written using MPLAB v. 7.43 (shown in Figure 5), the MPLAB C18 compiler v. 3.02, the MPLINK linker v. 4.02, and the TCP/IP stack used for the example was version 3.60. The TCP/IP stack is available as a free download from the Microchip website, and the example program described below is included.

The case study was developed around the project workspace file C18ProgramMem.mcw, in the directory where the downloaded TCP/IP stack was installed. This project is for the C18 compiler from Microchip, which is specifically targeted for the Microchip 8-bit PIC family (and is only licensed for Microchip products). If you have downloaded MPLAB and are taking a look at the project, take a look at the project display window, where you will see a hierarchy of all the files in the project, including C, H, and ASM files. If you are using the HPC Explorer board with the PIC18F8722 microcontroller, make sure under "Linker Scripts" you have the file "18f8722i.lkr" so the linker knows what target to compile to. You can try out the sample if you want, by default it will need to be connected to a network

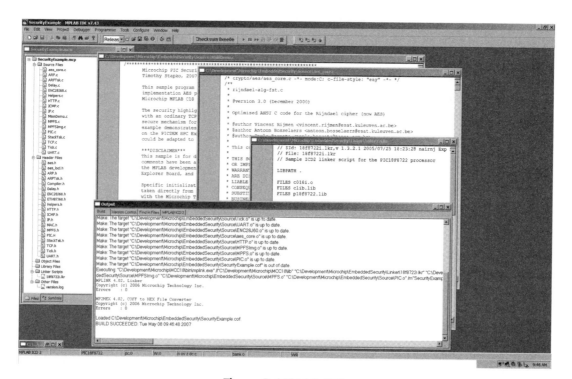

Figure 5: MPLAB IDE

with an active DHCP server, but that can be disabled in favor of a fixed IP address (what the example uses). Most of the user-configurable options for the TCP/IP stack are in the header file "StackTsk.h"—there are a number of macros with decent comments describing what they are for. The different pieces of the stack can be enabled or disabled by commenting (or un-commenting) the macros in this file.

The biggest hog of program memory resources in the HTTP server are the images and web pages themselves. The TCP/IP stack comes with a rudimentary file system called MPFS and a utility program that generates C constants to import data from files. We do not use the HTML interface in the case study, so the MPFSimage.c file should be rebuilt with an empty directory. The skeleton of the HTTP server is left intact in case the reader wants to try adding it back in, but it is not strictly necessary for functionality. For some reason, however, completely removing the HTTP server files from the project resulted in the hardware being unresponsive to Ethernet connection requests, but leaving it in does not use up too much extra space, so it remains.

Other than HTTP, the other network protocols can be completely eliminated—in StackTsk.h, all of the following macros can be commented out to remove pieces of the stack to free up some room for AES: STACK_USE_SLIP, STACK_USE_IP_GLEANING, STACK_ USE_DHCP, STACK_USE_FTP_SERVER, STACK_USE_SNMP_SERVER, STACK_ USE_TFTP_CLIENT, and STACK_USE_ANNOUNCE. Several of these macros have accompanying .C and .H files that should be removed from the project, and Microchip has done a good job of adding in-line macro error messages for the files that are no longer needed once the above macros are commented out. None of the protocols in the above list is needed by the case study example, but they can definitely be added back in (DHCP is particularly nice to have). Try experimenting with removing and enabling different protocols to get an idea of what will fit. Be careful, though, since if you push the code size past the 64 KB boundary, any constants beyond that boundary (above address 0xFFFF) will need to be declared "far." The C18 compiler does not warn you that your constants have been pushed into far space, which can lead to some frustration (the author spent many hours wondering why the AES password looked like HTML tags. The address of the password was far, but it truncated to a near address where some HTML had been stored). If you own the full version of C18 (this example was developed using the free feature-limited version), then you can just switch to the large memory model, which makes everything far.

The sample application itself is designed to be very general, so some of the extras (such as the LCD display driver) can easily be removed. Fortunately, the folks at Microchip did a pretty good job of protecting the files with macros, so the compiler will squawk at you if you forgot to remove something or removed something you shouldn't. Make sure at the end that you can compile everything (the main program file will likely need some modification).

Once the remnants of the HTTP sample are removed, then we can get down to implementing our secure PIC-based application, starting with the port of AES.

As we discussed earlier, AES is a block cipher and may be used as a stream cipher if some type of chaining or feedback is used (such as Cipher Feedback (CFB) or Cipher Block Chaining (CBC)). For our example, we are going to use AES as a block cipher. This usually entails adding and removing padding from messages that are not an even multiple of the block size (the AES block size for a 128-bit key is 16 bytes), but since we designed our command packet to be exactly 16 bytes, there is no padding required (the random bits added to the packet structure could be considered padding, as well as the reserved bytes). The AES implementation used for the example is from OpenSSL version 0.9.8d, which is originally derived from the publicly-available AES reference implementation in C. As we will see next, the porting to the PIC and the C18 compiler is fairly straightforward.

AES has a simple user API that will allow you to encrypt an arbitrary amount of data. There are a number of API functions that we could talk about, but we will limit our discussion to the following four:

- *AES_set_encrypt_key*: Takes a pointer to a key data structure and the key (an array of bytes), and the length of the key in bits (in our case, 128 bits).

- *AES_set_decrypt_key*: Same as AES_set_encrypt_key, but for the decryption side of the operation.

- *AES_encrypt*: takes a buffer containing the message, a pointer to an output buffer, and a pointer to a key data structure containing the information for the encryption key. The output (in the output buffer) is the encrypted message.

- *AES_decrypt*: The same as AES_encrypt, but the input is an encrypted message, the key is the decryption key, and the output is a plaintext message.

The basic application uses the TCP socket to transmit messages all encrypted using the AES algorithm. The application will use a predefined AES key that can later be changed. The initial key is included in the program and loaded to the device with the code (aes_password). During the first communication with the device, the password should immediately be changed (one of the capabilities of the program is to change AES keys remotely).

Porting AES over to the PIC involves copying three files from the OpenSSL/Crypto/AES directory into your MPLAB project directory. The files, aes_core.c, aes.h, and aes_locl.h should be moved into the appropriate source and include directories, and then added to the MPLAB project hierarchy. First, open up aes_locl.h. Locate the line with "#include <openssl/e_os2.h>" and comment out or remove the line. The other things to notice are the macros GETU32 and PUTU32, and the typedef of "u32." The PIC18F8722 has an integer size of 16 bits and AES needs a 32-bit type, so the AES_LONG macro must be defined (before the #include in the main program), or simply comment out the line with "typedef unsigned int u32," leaving only the long version. One other thing to note in the file is the "#undef FULL_UNROLL." Defining the FULL_UNROLL macro will result in a loop-unrolled version of AES being compiled, which may be faster but uses more code (this specific feature was not tested as part of the case study).

After saving and closing aes_locl.h, open up aes.h. This is the actual file that will need to be included in your main program (maindemo.c in the example). Locate the line with "#include <openssl/opensslconf.h>" and comment it out or remove it (it's not needed). The only other thing to note in the file is the AES key structure, typedef'd to "AES_KEY." Note the use of the AES_LONG macro in the key structure definition. Once you are done, save the file and open aes_core.c.

The entire AES algorithm is implemented in the aes_core.c implementation file. There are several modifications in this file—first find the line with "#include <assert.h>" and remove it. You will need to add a line with "#define assert(x)" to define the assert macro to nothing (you can go ahead and use assert if you want, but you will need to change the #include to point to the PIC version of assert.h). Also near the top of the file, you will need to locate the two lines that include aes.h and aes_locl.h. Remove what is there and add '#nclude "..\Include\aes.h"' and '#include "..\Include\aes_locl.h."' Next, you will see the AES lookup tables. The PIC stores constants in the program memory, and requires a special keyword, "rom" to indicate the location the constants belong. The lookup tables are named *Te0-Te3*, *Td0-Td4*, and *rcon*. Locate each definition and change "static" to "rom." If your data is past 64 KB and you are not using the large program memory model, do not forget to add "far" after "rom" (this will likely generate warnings in the AES functions—do not ignore these warnings!). There are no modifications to most of the functions, but there is an important one to make in AES_decrypt. At the very end of the function (which happens to be the last one in the file, around line 1136), there is a section with a comment indicating that the code performs the final round and maps "cipher state to array block." The C18 compiler does not up-cast the Td4 lookup table entries to a long, but the code applies left shifts that put the bytes in the upper half of a 32-bit quantity (variables s0-s3 have 32-bit types). You need to add a cast to the "u32" type to each expression containing a reference to Td4. The following is a before and after example (note the extra parentheses in the "after"):

- Before: `(Td4[(t0 >> 24) ])<< 24) ^`

- After: `(((u32)Td4[(t0 >> 24)]) << 24) ^`

There should be a total of 16 lines changed in AES_decrypt. Once the modifications are complete, save and close aes_core.c.

The Application

Now we can discuss the PIC application itself. A complete listing of the main program is provided in Appendix A, so the reader can follow along if desired. We will go through the example a section at a time, discussing the design decisions that affect the security of the application, and how the AES cipher is used in the application. While the primary focus of the discussion is obviously the use of AES, we will also cover some implementation details that are important to the overall security of the application—remember that security is like a chain, if there is one weak link, it all breaks down.

The first item of interest in the example is the AES key, stored in a program memory variable creatively called *aes_password*. The declaration of aes_password is as follows:

```
const rom far char aes_password[] = "PasswordPassword";
```

Note that the variable is declared as a "rom" variable in far space. The far qualifier is needed only if the declaration is beyond the address 0xFFFF, the top address that is reachable by 16-bit pointers. Also note that the "password" used as the key is exactly the 16 bytes required for an AES key. This key is very simple and should only be used at startup, since it is permanently stored on the device. A dump of program memory will reveal the key, so it would be good to protect the end device by disabling the programming interface in the final product (and maybe encase the entire chip in epoxy to make it really hard to get at). When we get into the main program, you will see that we need to copy the key from program memory into a RAM buffer, since the AES libraries are setup to only take pointers to RAM. This difference is important since the PIC architecture separates the instructions and data into separate, isolated spaces—the pointers are not directly interchangeable without some help from specialized library functions (or do the copy by hand using raw pointers, but this is prone to error).

Moving through the example program, note the declaration of the AES keys:

```
#pragma udata MAIN_MEM // This pragma enables our new memory area

    // Define 2 keys, one for encrypting outgoing messages and

    // a second for decrypting incoming messages. Note that we

    // could re-use the same key, but it is easier if we have 2,

    // so we don't have to keep clearing and reinitializing one.

    static AES_KEY en_key, dec_key;

#pragma udata           // Return to the existing memory area
```

The AES keys are quite large for the data space on the PIC (around 200 bytes per key—the PIC has a little less than 4 KB of RAM). For this reason, we actually need to instruct the linker that we need some additional space above and beyond what is normally allocated to the device. The way we do this is by modifying the 18F8722i.lkr linker script. Open the script, and locate the section where there are a number of DATABANK definitions (names are something like gpr0, gpr1, etc.). Choosing which banks to modify depends on the application, but most of them are allocated to various parts of the application by the linker. For the example, the databanks that are modified are gpr7 and gpr8. The original definition for these banks looks like this:

```
DATABANK      NAME=gpr7        START=0x700        END=0x7FF

DATABANK      NAME=gpr8        START=0x800        END=0x8FF
```

We need to adjust the banks to create a bank large enough to hold our key data structures. We basically steal space from one bank and give it to another, since we have a fixed amount of space to work with. In this case, we reduce the size of gpr7 and increase the size of gpr8. In the excerpt below, note that the new size of gpr8 is now 0x1EF (495) bytes, large enough to hold 2 AES keys (you could optimize the space by calculating exactly how many bytes are needed by the keys and adjust the banks accordingly). The gpr7 bank now only has 15 bytes of space, but it appears to be fine (the linker does not complain). Applying the changes results in the following definitions:

```
DATABANK      NAME=gpr7        START=0x700        END=0x70F

DATABANK      NAME=gpr8        START=0x710        END=0x8FF
```

Now that the linker script is modified, we need to add one more line to the script to define a space for our keys:

```
SECTION       NAME=MAIN_MEM    RAM=gpr8
```

Going back to the declaration of our keys, you will see that the first #pragma applies our newly created space MAIN_MEM.

For the most part, the rest of the declarations are straightforward, so let's skip forward to where we actually setup the AES keys. In the main function, you will see the following 2 lines:

```
memcpypgm2ram(key_buf, aes_password, 16);

AES_set_decrypt_key(key_buf, 128, &dec_key);
```

Note that we use a special version of the standard C memcpy, called memcpypgm2ram. This function specifically copies data out of the program space (in this case, our AES key password data) into a RAM buffer (a temporary buffer allocated in main). The next line in the excerpt sets up the decryption key that will be used for incoming commands. Note that it would probably be a good idea to clear out the temporary buffer after the key is set up (you do not want too many copies of the key hanging around), but since the key is already stored permanently on the device, it's probably not a big deal.

The primary point of interest in main() is the actual decryption of the incoming message. In the excerpt below, the encrypted message has already been received and is stored in the buffer input_buf. Note that after the decryption, the plaintext command packet will be in the output_bug buffer, which is then passed to the command processing function.

```
// Now that we have received a command, decrypt it
// and process it. The decrypted message will be
// in output_buf when we are done
AES_decrypt(input_buf, output_buf, &dec_key);


// Process the incoming command
ProcessCommand(output_buf);
```

Note that the format of the AES API is a little different than standard C, the destination buffer is actually the *second* parameter.

Finally, at the end of the main program loop is the code that sends a response to the PC client program:

```
memcpypypgm2ram(input_buf, aes_message, 16);
AES_encrypt(input_buf, output_buf, &en_key);
```

Note that the current implementation simply copies the static message from program memory into a RAM buffer and encrypts it. This is exactly the situation that was discussed above, and where the additional construction of the "response packet" should occur.

The final point of interest in the PIC program is the ProcessCommand function. After a command packet has been received and decrypted, it is passed into this function. Most of the code is pretty straightforward, but be sure to note the counter and checksum checks—the code checks the checksum first. To calculate the checksum, we simply add all the bytes in the packet together (other than the checksum itself), restricting the result to 8 bits. The result of the calculation is compared to the checksum byte in the packet. If they do not match, the packet is in error and nothing is done. The checksum assures that the random data has some effect on the security of the packet—without it, the random data is essentially padding, which isn't bad, but the checksum adds a little more "feel-good factor." A similar check is done on the counter value with the same result if a discrepancy is found. Note that the counter is not incremented if the packet fails—we don't want any state to change if an error occurs. There is a small synchronization problem if the error is caused by a legitimate packet that gets mangled in transit, but the problem can be resolved when no response is received by the PC client. The client counter should just be reset to the value it was after the last successful command execution.

Finally, we will look at the key change protocol in ProcessCommand. The function is essentially a state machine that allows for a secure key exchange to occur. The simple key-change protocol consists of 3 states, a "normal" state, a "key receive" state, and a "checksum" state, represented by the enumeration values STATE_NORMAL, STATE_RECVKEY, and STATE_CHKSUM, respectively. In STATE_NORMAL, commands are processed as usual, and the LED states are changed appropriately. However, if the command byte has the "change key" bit set, the state machine is moved into the STATE_RECVKEY state to prepare for receiving the AES key. The AES key we are about to receive is exactly the same size as a command packet (16 bytes), so we need the state machine to be sure we receive it correctly. When in the STATE_RECVKEY state, the next pass through the main program loop will receive the encrypted key and pass the decrypted plaintext key into ProcessCommand. The key is copied into a temporary buffer, and then the state machine is moved to the STATE_CHKSUM state. Note that we do *not* update the key at this point. After the key is saved, the program returns to the main program loop to receive the third packet in the key change protocol sequence. This third packet simply contains a checksum of the key sent in the previous packet. When the plaintext checksum is passed into the ProcessCommand function, the final state is entered. If the checksum checks out (compared to the temporary key), then the decryption key is updated. Finally, even if the checksum failed, the state machine is returned to STATE_NORMAL to be ready to receive command packets as usual.

The reason for the change-key protocol and the checksum is that the second packet containing the key could be corrupted in transit (or modified by an attacker). If we simply updated our key upon receiving that packet, we have no way of knowing whether it is the real key or random data. If the packet was corrupted, the resulting data could be used as an AES key, but the client program would have no way to communicate with the device anymore (decrypting the corrupted data would result in essentially random data that would then be used as a key). Sending the checksum in a third packet assures that the key received in the STATE_RECVKEY state was the intended key.

We have so far just looked at the PIC side of the application, but we need a client application in order to use the hardware. In this case, we do not have the luxury of being able to use a preexisting client like a web browser, so we have to build our own. In the next section, we will look at the PC-based client application that we will use to connect to our PIC over a network.

The PC Client

With the PIC-side of the application complete, we are now ready to implement a PC-based client to communicate with our hardware. For our case study, the client program was implemented using gcc 3.4.4 for the Cygwin environment (a Linux-style shell emulator for Windows—see Figure 6), and the AES implementation from OpenSSL 0.9.8d. The client program's purpose is to provide a human interface for the command protocol we developed for this application. The program (full listing of the main program is in Appendix A) consists of a simple text-based command-line menu interface based on standard C I/O functions (printf and scanf). We are primarily concerned with how the security of the example operates, so the interface could definitely use some improvements, but it is sufficient to demonstrate the capabilities of the PIC application.

Writing an AES application for a PC is a much easier task than for the PIC, so we will not go into as much detail on how AES was brought into the application from OpenSSL (the AES code essentially required no modifications). The application itself is pretty simple—it takes the user input from the menu interface, and constructs a command packet. The user has the options to change LED states, set the "LED Preserve State" command bit, reset the

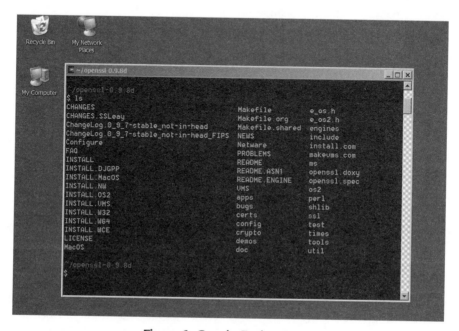

Figure 6: Cygwin Environment

global counter, and change AES keys—all of the options provided by the command protocol implemented on the PIC side. While the program is running, the global counter is updated as each command is sent, and the command packet itself can be modified and resent to the target. The user also has the option to reset the command packet and print the current contents of the command data structure (which is simply an array of bytes that is parsed by the print routine).

In the client program the key change protocol is simply implemented using a flag that indicates that the key is to be changed. Upon entering a key change command, the flag is set, and when the user sends the command packet, the program requests that the user enter the new key. The application then sends the key packet (encrypted, of course), and prompts the user about changing the key. If the user requests the key change, then the checksum packet is sent.

The client program is a very rudimentary interface for the device, but it illustrates the basic protocol, upon which a better application could be built. Now, with our brief discussion about the client program done, let's look at what we learned and what improvements could be made to the PIC application.

PIC Example Wrap-Up

So what have we learned with the PIC example? First, security is more than just encryption. As can be seen in the command protocol that we developed, the encryption is just a building block upon which security is built. The inclusion of random data, the counter, and checksums are what make the encryption effective. Without varying the data contained within the encrypted packets, an attacker could easily build a library of messages for replay attacks. Even the simple response mechanism, which seemed innocuous enough, could be the target of an attacker—even something that seems like a good idea can result in an unintended vulnerability.

The PIC case study was also an example of how security can be designed into even the most resource-constrained systems. The other case studies are more about using the security options provided with the hardware, or complete security solutions such as SSL. What we have also illustrated with the PIC is that you can build your own security into your application, but you have to be extremely careful. The command response mechanism was purposefully left vulnerable to illustrate this fact. The requirements and even the design of the application may look secure, but if you don't think of all the possibilities (think like the enemy), it doesn't matter how secure the rest of your application is, because the piece you forgot will be the piece that is successfully attacked.

Now that we have looked at the PIC example, there are a few improvements that could be made to the design if it were to become a real product. First, as was noted already, the response mechanism would benefit from some additional security measures to help prevent the man-in-the-middle vulnerability. A second improvement would be to add some additional security to the third packet containing the checksum in the key change protocol. Currently, the checksum is a single byte, and though it is somewhat unlikely that the checksum would result in a false positive (indicating the key is valid when it actually is not), there is a chance it could happen. A larger checksum or, even better, a cryptographic hash could be applied for verification of the key. It might seem that resending the key in a second packet might be sufficient to improve the security from the one byte checksum, but remember that two packets containing the same data will result in the same ciphertext, information an attacker may be able to use. Additionally, resending the same data could actually increase the possibility of a random 16-byte string being used as the key accidentally (think of a network gone awry, sending the same random packets out—although unlikely, two of the same packet in succession at the right time would be enough to create a problem). Finally, there is the issue of the key being stored permanently in the device. The best way to combat this is to change the key upon deployment of the device to something stored in volatile memory. What would possibly be even better would be to allow the permanently-stored key to be updated periodically. The PIC provides some EEPROM memory that may be useful for that purpose, since it can be rewritten by the application.

CHAPTER 12

Rabbit Case Study

Rabbit 4000 CPU with Dynamic C

Before we discuss the Rabbit microprocessor, we need a disclaimer. The author, at the time of this writing, is employed by Rabbit Semiconductor (who makes the Rabbit processor) and its parent company, Digi International. The Rabbit is a very popular and powerful embedded systems platform, so it deserves a special place in this book, especially with some of the security features that Rabbit has incorporated into its processor design. However, due to the author's affiliation with Rabbit, it is only proper to warn you, the reader, that this section will be biased toward the positives of the Rabbit. That being said, all attempts will be made to view the Rabbit in an objective light, covering both the positive and negative aspects of the hardware and software. Now, with *that* out of the way, let's take a look at the Rabbit.

The History of Rabbit

The Rabbit is a relatively new addition to the family of 8-bit processors that share some design similarities with the venerable Z80 from Zilog. The Rabbit was specifically designed to overcome some of the limitations of the Z80 and its brethren, while still retaining the basic feel—a CISC instruction set with 8-bit functionality. The Rabbit added a number of more modern features, including better stack support and instructions designed to support compiler-generated assembly from the C programming language. The Rabbit also added better 16-bit memory support and a segmented memory model to support up to a total of 1 megabyte of memory. The Rabbit, as originally designed, was to be part of a larger system, with individual chip sales as a secondary market. The Rabbit was introduced in 1999 as the Rabbit 2000, and it was primarily sold in single-board computers and what are known as *core modules* manufactured by Rabbit's parent company Z-World, Inc.[1] A core module contains the basic circuitry, memory, and

---

[1] In May 2005, Rabbit Semiconductor and Z-World, Inc. were purchased by Digi International, a leading manufacturer of embedded networking hardware. As part of the acquisition, the Z-World brand was dropped in favor of the more recognizable Rabbit brand.

Figure 1: Rabbit Core Module and Development Board

interfaces needed to utilize the Rabbit processor's capabilities. A typical core module from Rabbit contains a flash memory for storing program code and data, an SRAM or multiple SRAM's for execution, and a programming header (designed to work with a special serial cable). Newer variants of the core module include analog input, Ethernet support, and mass storage in the form of NAND or serial flash.

The Rabbit is now in its third generation, the Rabbit 4000, which is the processor we cover in this book. The Rabbit 4000 represents a new direction for the Rabbit with nearly twice the instruction set of the Rabbit 2000 and 3000 processors, and support for up to 16 megabytes of program memory. The Rabbit 4000 also introduces some security features actually built into the silicon, which are especially interesting for our discussions. As we will see, the Rabbit has hardware support for speeding up public-key operations such as RSA, and includes some built-in instructions for hardware-assisted AES. Another attractive feature of the Rabbit 4000 is the inclusion of a full 10-base-T Ethernet controller into the silicon of the processor. All that is needed to support an Ethernet connection are the physical networking components and an external interface to the RJ-45 connector. Several core modules feature this circuitry pre-built to use the internal controller, which saves significant cost and circuit board real-estate for Ethernet-enabled core modules.

Software on the Rabbit

From its introduction in 1999, the Rabbit has been supported by a complete development environment called Dynamic C that includes a C compiler (with some language extensions we will discuss later), an IDE with a code editor and debugger, and an enormous collection

Figure 2: Rabbit RCM4010 Development Kit

of drivers and sample programs. Dynamic C provides everything a developer needs to develop the software for a Rabbit application, but some engineers may be turned off by some of its quirks. The Dynamic C language is not currently ANSI/ISO C compliant, though some strides have been made to make the basic language function more like a compliant implementation (despite many web forum discussions to the contrary, Dynamic C is really C, its just a different flavor). Also, Dynamic C does not support a full traditional linker model, so an application must be compiled all at once, rather than in steps as with a more traditional compiler/linker model. Despite these shortcomings, Dynamic C is a popular and powerful development environment that is in use in thousands of successful applications, and you can't beat the cost—it is priced significantly less than many other professional development tools (except for Open Source tools like the gcc compiler suite),[2] and it comes bundled with Rabbit hardware in inexpensive development kits (which sometimes cost less than a software license for Dynamic C).

For our sample application, we will focus on developing with Dynamic C, since it is the most common software platform used on the Rabbit. However, for those wishing to use something other than Dynamic C, there are some third-party options, such as the Softools compiler, which is a more traditional professional development environment with an ANSI/ISO C compiler and a traditional linker. For more information about the Rabbit, Softools, and other third-party options, please refer to the primary reference on developing applications with the Rabbit, *Embedded Systems Design using the Rabbit 3000 Microprocessor*, by Kamal Hyder and Bob Perrin.

---

[2] With gcc you don't get tech support, so the actual cost incurred through the learning process may be significantly more.

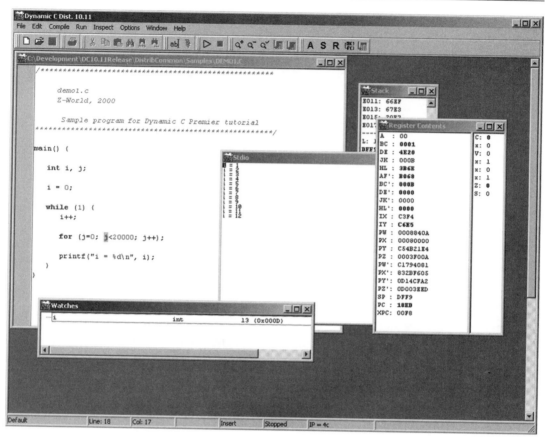

Figure 3: Dynamic C IDE

Rabbit Case Study—Internet Enabled Vending Machine

Given the hardware and software available in the development kit, we can look at developing a relatively feature-rich application. However, in keeping with the spirit of this book, we will only include those features that are absolutely necessary for the application to meet its needs. Resist the urge to add every bell and whistle to your application. Remember that each additional feature is a multiplier on the complexity (and therefore insecurity) of your application!

For the Rabbit, we will look at an Internet-enabled beverage vending machine.[3] We want to have several basic features in our implementation that make the application useful:

---

[3] The vending machine example is probably overused and definitely quaint, but it gets the point across.

- The vending machine should be able to communicate securely over the public Internet.

- The hardware in the vending machine must monitor several variables and report to a central server when certain criteria are met.

- The vending machine, being on the Internet, should have remote configuration capability.

The above list is the management-level view of the application. Let's take each bullet-point and dissect it for technical details. The following discussion will constitute the design for our application.

First, let's look at the criteria for sending a report or alert to the central server:

- For all alerts and reports, a configurable delay will be invoked once the message is sent successfully (which means we need to acknowledge messages received by the server). The delay is intended to prevent a denial-of-service condition (intentional or not), so a hard-coded minimum of 15 minutes will be enforced.

- An alert should be sent if the temperature ever rises beyond a configurable level, indicating failure of the refrigeration unit. For security reasons, we will hard-code a maximum configurable value and have an internal timer that will shut down the machine automatically if no actions are taken after the temperature alert level is met and a certain amount of time has passed.

- The machine should periodically report the number of beverages remaining, and the number of each type of beverage (i.e., cola, lemon-lime, and root beer). This number is not necessarily useful to an attacker but could be, so we will be sure to encrypt the report.

- If the beverage count ever reaches a configurable minimum, we will immediately send an alert for the machine to be refilled. To prevent an attacker from benefiting from this information, not only will the alert be encrypted, but we will hard-code a maximum value for the alert level (this prevents an attacker from causing the machine to constantly send alerts).

- The proceeds the machine takes in will also be reported with the beverage count. This number may be useful to an attacker who wants to rob a number of machines. To maximize the theft, the attacker could benefit from only robbing machines that are full of money. For this reason, we will definitely encrypt the report.

- Additionally, we will send an alert if the amount of money in the machine reaches a certain level. This is to back up the beverage count alert in case there is ever a situation where there is more money or more beverages than makes sense. If the number of beverages sold does not match the proceeds, we either have a bug or a security breach—in either case, someone should probably take a closer look.

Now let's take a look at the requirement that the application be able to communicate over the Internet securely:

- We have both SSL and AES available to us on the Rabbit. Unfortunately, the current implementation of Rabbit SSL is only available to us through the web server as HTTPS. For this reason, we will use AES for sending reports and SSL (via HTTPS) for the remote configuration requirement.

- To send a report, we will simply use a plain TCP/IP socket to connect to the server. However, the information that will be sent to the server will be encrypted using AES before being sent. The message will then be decrypted by the server upon receipt.

- AES requires a key to be known by both the central server and the Rabbit. We will generate keys at the central server and send them to the Rabbit over HTTPS. See the remote configuration breakdown (next) to see how we will handle this.

- As an additional security measure, we will utilize a special feature of the Rabbit 4000, which is an internal bit of SRAM (32 bytes) that is designed to hold cryptographic keys. If someone attempts to access this "register" via the programming interface on the Rabbit (the only way to get at the contents of the memory), the entire 32 bytes is cleared, destroying the key. This will prevent an attacker from being able to get to the AES key even if he has physical access to the hardware.

Finally, let's look at the remote configuration requirement:

- As we mentioned above, we will be using SSL via HTTPS for secure remote configuration. The Rabbit software libraries include a full-featured HTTP server that we will utilize to provide a simple but effective secure web interface to the vending machine.

- The Rabbit will need an SSL certificate. Fortunately, the SSL software bundle (which Rabbit calls a software "module") comes with a utility to generate the certificate, which makes it easy to do.

- We will send the AES key to the Rabbit using the HTTPS configuration page with a simple HTML entry field in the configuration webpage. We could improve on this by making the server update the keys automatically, but this would require some

HTTPS hacking (grab a "special" webpage designed for the server and automatically post the key). To keep the example simple, however, we will stick to a simple interactive webpage where the key is entered by the user (as an exercise, how could this compromise the security of the system?).

Now that we have dissected the application into its technical requirements, we can start looking at how to implement the application and apply the security concepts we have learned to the security problems we have identified in our design.

The implementation differs slightly from the application description to focus on the specific security features in which we are interested. Most notably, we focus on the SSL interface and setting up the AES key and initialization vector, and the protocol we develop to send alerts via plain TCP/IP (see the listing in Appendix A). On the Rabbit side, we include a simple mechanism that uses a pushbutton on the RCM4010 development board to send alerts at will (our alerts are just simple messages that get printed out by the server on the PC). This same mechanism could be modified to respond to anything the application may do, and the format of the packets allows for arbitrary-length data to be sent. There are numerous ways to set up the data that will be sent, but since we are focusing on the security mechanisms, we purposefully omit those implementation details. On the PC side (using the Cygwin environment with gcc for Cygwin/Windows), the application sets up a simple TCP/IP server that the Rabbit will connect to when sending alerts. It prompts the user for an AES key and initialization vector (these should match the hashed AES password and initialization vector printed out by the Rabbit program), and processes a single message. The PC sample could be modified to be a resident server, but we are primarily interested in the Rabbit code, so it has been purposefully left simple.

Collecting the Pieces

The Rabbit is a fairly powerful embedded platform, with enough horsepower and memory capability to support running a full web server along with a user application. At nearly 60 MHz, the Rabbit 4000 is similar in processing capability to the original Intel Pentium processor, in a form factor many times smaller. In fact, some core modules have a significantly smaller footprint than the packaging some Pentium processors came in (see Figure 4). The Rabbit 4000 used for our application came in a development kit (specifically, the RCM4000 Development Kit, graciously donated by Rabbit Semiconductor) that includes the Rabbit in a core module with Ethernet (the core module model is actually the RCM4010, which comes with 512 KB each of parallel flash and SRAM, along with the Ethernet circuitry), a prototyping board that includes a power supply for the core and has some space for building additional circuits, a programming cable (a serial cable for programming the core module), a serial cable for debugging, an Ethernet crossover cable for development, a selection of jumpers and standoffs, a quick-start manual, and the Dynamic C development environment (in this case, version 10.05) on CD-ROM. Rabbit was also kind

Figure 4: RCM4010 and Pentium III

enough to donate some additional software libraries (sold separately) for both SSL/TLS and AES support.

Setting up the Rabbit is a very easy task, simply load the Dynamic C software, plug the core into the prototyping board, plug in the power supply (a wall-wart transformer with a regulator on the prototyping board), plug in the programming cable, and attach the programming cable to the header on the core module. Additionally, the crossover cable may be plugged directly into your PC's Ethernet jack and into the Rabbit. Upon starting Dynamic C, a sample program may be loaded and compiled directly to the target. The software automatically recognizes the core module type and processor, and compiles the correct code accordingly. The program can then be executed and debugged directly on the Rabbit hardware. Neat!

Now let's focus on implementing our design using Dynamic C. First, let's take a look at some of the sample programs supplied with the software to give us a basis for developing our application. Since we are planning on using HTTPS, let's first take a look at the HTTPS samples, and specifically a sample that uses an HTML form with entry fields. We will also look at a basic TCP/IP sample and an AES sample for the other parts of our application.

We are implementing a web server with SSL capability, so the first samples to look at should be relevant to that goal. Fortunately Dynamic C comes standard with several hundred sample programs that demonstrate the capabilities of the different included libraries. We are first going to take a look at HTTP forms and using them over an HTTPS (SSL-secures HTTP) connection. For our remote configuration, we want to be able to control the user-modifiable options defined in our requirements, as well as generate reports on demand and provide an ability to update the symmetric AES key when needed.

The sample program we will look at first is called *ssl_form.c*. The program generates a simple form that demonstrates a web interface for a fictional thermostat application. Since our vending machine has temperature control as one of its requirements, this sample is a perfect starting point for our software. The key points to take from the sample are the SSL configuration and the basic setup of a web server, both requirements for our case study. Let's take a look at each of these features in a little more detail.

The Dynamic C web server libraries are controlled through compile-time macro definitions that give the user control over what features are compiled into his or her application. The following macro definitions control the number of servers in the application:

```
#define HTTP_MAXSERVERS 2

#define MAX_TCP_SOCKET_BUFFERS 2
```

The first macro, HTTP_MAXSERVERS, indicates the number of HTTP server state data structures to allocate—each one can be configured to provide different HTTP services and spread the load of incoming traffic. Most applications, including our case study, should only need one or two. The second macro indicates the number of buffers to use for TCP socket communications. More buffers allows for more simultaneous connections, but obviously uses up more data space.

Dynamic C, as was mentioned, is not ANSI C, so including a "library" compilation unit is a little different. The following line serves to bring the HTTP server functionality into the application:

```
#use "http.lib"
```

That's it. Now we can start using the web server functions. This case study is all about the security, so from here on, we will only cover the aspects of the program that touch on or are affected by the security pieces of the application.

To implement our application, we will start with the Secure Sockets Layer. SSL does not come standard with the Dynamic C software, but it can be purchased as a separate set of libraries (the same goes for AES). Most Dynamic C libraries are simply provided as source code libraries that are formatted to be imported into your application using the #use language feature mentioned above. At this time, however, the SSL libraries are only provided in encrypted format. While this will prevent us from using SSL sockets directly, our application is designed with this in mind, since we will be using AES for most of our communications, and using SSL just for the web interface and updating the AES keys. The SSL implementation is designed to work within the context of the HTTP web server, so we should not need to do anything "under-the-hood" with SSL (you probably would not want to anyway, unless you are curious about how SSL works).[4] The magic code that enables SSL in our sample program is a couple of macro defines that tell the HTTP server to include the SSL code and set up a server instance to use the secure socket:

```
#define HTTP_SSL_SOCKETS 1

#define USE_HTTP_SSL
```

The first macro, HTTP_SSL_SOCKETS, indicates the number of SSL sockets we want to dedicate to HTTP servers. If desired, we can allow both HTTPS and HTTP servers to be resident. The number of plain HTTP servers is determined by subtracting HTTP_SSL_SOCKETS from HTTP_MAXSERVERS. In our sample program, we have allocated 2 socket buffers for 2 separate servers, one secured with SSL, the other being a plain HTTP server. The second macro simply causes the SSL code to be linked into the program.

Before we go any further, we need to create an SSL certificate. Fortunately, the Dynamic C SSL software bundle comes with a utility that provides an easy-to-use wizard-style interface that makes creating a certificate quick and easy. Once the wizard is complete, the utility exports a file containing the certificate that is formatted to work with the SSL libraries. This file needs to be imported into our program (and downloaded to the Rabbit hardware when the program is compiled), so we will use another Dynamic C feature, #ximport:

```
#ximport "cert\mycert.dcc" SSL_CERTIFICATE
```

The "x" in ximport simply stands for "extended memory," and the file "mycert.dcc" is the output from the certificate utility. The final part of the statement, SSL_CERTIFICATE, is a macro definition that points to the file (which will be downloaded into constant memory space with the application). In this case SSL_CERTIFICATE is a name recognized by the SSL libraries as representing the certificate to be used.

Now that the program is set up to run with SSL, there is one little detail that needs to be handled at runtime, reserving the TCP port we need for the HTTPS connection. In the Dynamic C TCP implementation, reserving a port allows multiple connections to be accepted without having to use additional resources for each new connection. This nifty trick is done with the following function call:

```
tcp_reserveport(443);
```

As was mentioned when we discussed SSL earlier, TCP port 443 is the reserved port for HTTPS (like port 80 for plain HTTP).

---

[4] If you want to know the nitty-gritty details of how SSL works, check out the OpenSSL implementation (www.openssl.org), an open-source industry-standard implementation available free for download. Unfortunately, OpenSSL is too big for many embedded platforms, so proprietary implementations will have to suffice.

Now we have a working SSL web server application. We need something to protect with our newfound security. In the Dynamic C environment, there is an easy-to-use API that allows the programmer to create web forms and associate them with particular URLs and HTTP server instances. The form in our sample program is an HTML form represented by a form variable called *myform*. In the following line of code, the form (which we are creating in code) is associated with the URL "myform.html" and with the secure HTTPS server.

```
form = sspec_addform("myform.html", myform, 5, SERVER_HTTPS);
```

We are not going to delve into the details of creating the entire form here, but each field is added to the form using the sspec_* API. Each element of the form must also be associated with SERVER_HTTPS in order to be secured with SSL. See the application listing in Appendix A for more details.

The last detail that we need to cover in the HTTPS interface is the matter of password-protecting the Rabbit HTTPS server. We need to use some type of authentication since SSL only protects the communications channel. Without some form of authentication, such as requiring the client to enter a password, any client could just connect to the Rabbit and start modifying parameters! For the authentication, we can devise our own password mechanism or, even better, we can use either of the primary HTTP authentication methods, basic or digest. With HTTPS, the authentication occurs after the SSL connection is established, so we can do a basic HTTP authentication (which sends the password as plaintext over the network) without having to worry about eavesdroppers. However, we can also use the digest method if we are really feeling paranoid. There is not any real reason we need to use digest instead of basic, but it is always more reassuring to have multiple redundant levels of security, just in case. For this reason, we will enable HTTP authentication with the following macro definition:

```
#define USE_HTTP_DIGEST_AUTHENTICATION    1
```

This macro simply tells the HTTP server implementation that we want to use digest authentication so it will include the appropriate functions. To actually enable the authentication for the server, we need to look at the following code:

```
http_setauthentication(HTTP_DIGEST_AUTH);

admin = sauth_adduser("admin", "password", SERVER_HTTPS);

page1 = sspec_addxmemfile("/", index_html, SERVER_HTTPS);

sspec_adduser(page1, user1);
```

The first line of code in the snippet above enables the digest authentication mechanism in the HTTP server. In the second line, the call to *sauth_adduser* adds a user to the server that can later be associated with particular resources. In this case, the username is "admin" and the password is set to a default of "password," and the last parameter indicates that the user should be added to the HTTPS server, rather than just the plain HTTP server. The variable *admin* represents the user ID number for the user we just added. This variable will be used whenever we need to change the settings pertaining to the "admin" user (such as changing the password). In the third and fourth lines, we create a resource for the HTTPS server that is the root of the server's URL, associate that root with the *index_html* resource, and then restrict access to that resource to only the "admin" user. The Rabbit HTTP server uses an API that provides an abstraction layer for HTTP and HTTPS resources, which may be drawn from numerous sources. In this case, the root page is actually an imported file that resides on the flash (if you recall, in the HTTP form sample program, we created the index.html "file" at runtime—this is another option). With the resource associated with the user, any attempt to access that resource (in this case, the root URL of the Rabbit device) will prompt the browser user for a username and password. If the username is unrecognized or the password is incorrect, the server will prevent access to the resource. This forms the basis of our authentication for incoming users on the HTTPS interface.

With a working web interface, we now will turn our attention to the back-end details of the application, namely the AES encryption. The AES key will be imported from an HTTP form field and copied into the internal tamper-detection SRAM, which will hold up to 32 bytes. If the smode pin is toggled (indicating that someone is trying to access the internal processor state of the Rabbit via its programming port), the 32-byte register will be cleared, assuring that the attacker cannot retrieve the key. This is an added bit of protection that can go a long way if the hardware resides in a remote location and is susceptible to attempted physical attacks. Storing and retrieving the key is as simple as a memcpy operation, in this case, a special memcpy called *root2vbat*. In this case, *root* refers to the logical memory space of the Rabbit, and *vbat* refers to the internal register (also called *VBAT RAM*). Rabbit is kind enough to provide a sample demonstrating this feature in a program called *tamperdetection.c*, in the RCM4000 samples directory.

The AES API is pretty straightforward, so all we need to do is to combine our code for loading the key with the AES code. For this, we will look at an AES sample to see how we setup an AES encryption. There are several different options for using AES, both as a simple block cipher, or in a little more complex setup, where the block cipher is used as a stream cipher. To turn AES into a stream cipher, part of the ciphertext from each encryption operation is fed back into the algorithm, through what is called an *initialization vector*, or IV (sometimes also called a feedback register), which is a chunk of data the size of the cipher block that is recalculated after each operation. There are several ways the initialization vector can be set up, but in the Rabbit implementation, a standard scheme called *cipher feedback*, or CFB, is used. One important thing to note in this case is the fact that the encryption and decryption IV's are different, so a different state for each operation must be kept. In our application, we will keep one vector on the Rabbit for encrypting outgoing messages, and a second for the decryption of incoming messages from the server. AES stream encryption is demonstrated in the sample program *aes_streamtest.c*, in the AES_ Encryption samples folder in the Dynamic C installation.

To make it a little easier on the operator, we will allow the AES key to be entered as a phrase that is easier to remember than a random string of data. This will inherently decrease the security of our AES encryption, since the implementation will be more susceptible to a dictionary attack (selecting known phrases and words from a dictionary and trying each one until it decrypts the data correctly). However, since our application is a vending machine and not a nuclear bomb launch system, we do not have to worry as much about an attacker spending the time and resources to mount a large dictionary attack (if you *are* implementing a nuclear bomb launch system, then you should probably look for references on security in other places than this book—please!).[5] Assuming this decreased level of security is acceptable, we will use some more sample code to develop our key generation. The sample code uses a simple MD5 hash to take in an arbitrary string (our pass-phrase) and generate a fixed-length 128-bit AES key. MD5 just happens to output exactly 128 bits (16 bytes) per operation, so we don't even have to worry about generating enough key material (we could also use SHA-1, which generates 20 bytes, but all we would have to do in that case would be to extract the 16 bytes we need). The source code from the sample (aes_streamtest.c) that performs this bit of magic is as follows:

```
md5_init(&md5_state);

md5_append(&md5_state, passphrase, strlen(passphrase));

md5_finish(&md5_state, key);
```

The first function call to *md5_init* initializes the hash (the variable *md5_state* is an MD5 instance structure), and the second call to *md5_append* actually hashes the phrase into our key material. Finally, the third call to *md5_finish* puts the hash into the key array for our use. One important thing to note here is that the MD5 state structure now contains the information needed to get the key, as well as the *key* array. If we are going to the trouble of using the tamper detection internal RAM to protect against someone reading out the key, then we should also clear any memory that may contain that same key material. This clearing of memory will be shown in the next section when we put all these pieces together into an application.

Putting It All Together

Now that we have covered all the parts of our application, we can start to put together the parts to form our complete application. Even though we are taking advantage of the

---

[5] Not to mention that you should probably **not** be putting a nuclear weapons launch system on the public Internet!

extensive sample program collection that is supplied with Dynamic C for the Rabbit, it is a good habit to develop and test your application in such a modular fashion. If you start from small pieces, it is easier to predict the interactions of those parts when you finally put them all together.

The first thing we are going to look at is the AES key update logic, as this is a core component of the security of our application. The core of this logic revolves around an array that will contain the key material needed for the AES operation. In this application, we can actually take advantage of something that will make our lives easier. Use the same array to always store the key material, whatever format it is in, from the original pass-phrase text to the completed hash. By using a single array to hold the key, we gain two distinct benefits: one, we save valuable data space by using only one buffer, and two, we only have to keep track of one instance of the key. If we were to use multiple buffers, one for each step of the process of updating the key, then we would have to worry about all those copies of the key existing simultaneously, and we would be more likely to miss one when clearing memory (and it *will* be the one you missed that the attacker will use to defeat the security of your application).

To see what we now have in the final application, the following is a snippet of the key update:

```
// Initialize the MD5 hash
md5_init(&aes_md5_state);

// Add the AES passphrase from the textbox to the MD5 hash
md5_append(&aes_md5_state, aes_passphrase,
strlen(aes_passphrase));

// Finish the hash and put the result in the AES key buffer
md5_finish(&aes_md5_state, aes_key);

// Set up the new AES key
// Only use half of the 32 bytes available
root2vram(aes_key, 0, 16);

// Clear out all temporary data buffers associated with the AES
// passphrase and key generation. The key is now stored only in
// the tamper-detect RAM register
```

```
memset(&aes_passphrase, 0, sizeof(aes_passphrase));

memset(&aes_md5_state, 0, sizeof(aes_md5_state));

memset(&aes_key, 0, sizeof(aes_key));
```

In the above code, the array *key* is associated with the HTML field in the HTTP form (which is of course protected by SSL). The code is called when the user enters a new AES pass-phrase, and does the appropriate hashing with MD5 according to our design. Once the hash is complete, the key is copied into the internal RAM register (with the tamper-detection feature), and then ***the array is cleared.*** The last step is very important, as we do not want to leave the key lying around in external memory for an attacker to find. Since we only used the one array to represent all forms of the key, it is a simple matter to erase all evidence of the key from "normal" RAM. Note that this should be done at any time the key is copied out of the internal RAM and is done being used.

We can also use the other half (16 bytes) of VRAM to hold the password used to access the HTTPS server. This password must be entered by the user in HTTP authentication over HTTPS as we discussed above. The HTTP server will keep a copy of the password stored internally (unfortunately), but we will use the VRAM copy to verify the old password when the user wants to change to a new password. If desired, we could do some investigation into the source code of the HTTP server to find out where the password is stored and used for authentication. We could replace the logic in the HTTP server libraries (Dynamic C comes standard with all the source code for most features) that does the actual password comparison with some code that utilizes the VRAM copy instead of a copy resident in main memory. To do this, we would implement a temporary buffer to copy the stored password from VRAM, do the comparison, and then clear the buffer, leaving the only copy of the key in the VRAM register. This work depends on the versions of Dynamic C and the HTTP and TCP/IP libraries being used, so we will leave the investigation and actual implementation of this additional security measure as an exercise for the reader.

We setup the authentication using the following code, which will assign a new password entered by the user to the *admin* account and place a copy of that key into the VRAM register:

```
// Password has been submitted by user, stored in array "key"

if(strlen(key) <= 16) {
root2vram(key, 16, 16); // Use the other half of VRAM
sauth_setpassword(admin, key);
    }
    else {
```

```
                // Password entry error, leave password the same
    }
memset(key, 0, sizeof(key));
```

To save on space (and to assist in clearing all instances of the key and password), we are reusing the "key" array for the password as well. We need to be careful that the form limits the password to a maximum of 16 characters and indicate the overflow error to the user if the password entered is too long. We should also double check the length in code, rather than simply rely on the HTTP form to limit the input. As you can see from the code above, we also do a check after the user has submitted the password to be sure an overflow was not introduced in the interim. If we get this far and the key array containing the password is not a properly-formed null-terminated C string, we leave the password alone. In the "else" block, we can choose to put additional code to notify the user of the error, but this is a user-interface detail so we omit it here.

Now that we have looked at the user interface and how we will update the passwords and keys used by the application, it is time to take a look at the implementation of the communications layer that actually utilizes AES for securing transactions. Fortunately, this operation is as simple as opening a plain TCP/IP socket and encrypting any data that we send back to the PC server using that socket. In this application, we will use AES in stream-cipher mode, which simply means that we will use it so we can encrypt a stream of bytes, rather than a block at a time. Fortunately, Rabbit's AES implementation comes standard with a stream mode called Cipher Feedback, or CFB. This mode of AES takes some of the encrypted data from each block and feeds it back into the next AES operation as part of the "key." In CFB mode, the "key" used for every block (other than the very first block operation, which uses our real AES key) is constantly changing once the algorithm is started. For this reason, we need to keep the current key state around on both ends of the connection or we will be unable to decrypt incoming messages on the PC side, even if we have the real AES key. Fortunately, the Rabbit implementation of AES already has the state structure for this mode, and even provides an AES-CFB stream API. All we need to do is be sure that the PC implementation of AES also supports AES in CFB mode.

On the Rabbit, the code to do an AES encryption operation in CFB mode is very simple. The following snippet shows the setup of an AES stream and how it is used to encrypt outgoing data:

```
AESinitStream(&encrypt_state, key, init_vector);

memset(key, 0, sizeof(key));

// …
```

```
// Loop as needed, etc…

AESencryptStream(&encrypt_state, data, strlen(data));

// Send data here

// …

memset(&encrypt_state, 0, sizeof(AESstreamState));
```

The first line of code actually sets up the stream cipher, using the key (which in our application is our temporary key buffer), an *AESstreamState* variable, and *init_vector*, which is an AES initialization vector used to start off the stream (in our case, *init_vector* is an array of bytes equal to the AES block size initialized to some known values—1, 2, 3, . . .16, for a block size of 16). Following the initialization of the AES stream, we no longer need the copy of the key, so we clear it in the second line. Once the stream is initialized, we set up the TCP/IP socket for our communication (omitted here for clarity) and encrypt the data we want to send in the call to *AESencryptStream*. Note that the data is encrypted in place, replacing the plaintext data in the buffer *data* with the ciphertext. We continue to send data using our stream until we are done, then we clear out the *encrypt_state* structure to eliminate evidence that could be used to reconstruct our key. For simplicity, we will assume that we will only allow fixed-size records of data to be sent in this manner, so if the PC end does not receive a full record, then a truncation attack error will be flagged.

The PC Side

We have now implemented the Rabbit side of the application, but we need the server that the Rabbit will connect to for uploading information. This is a book on *embedded* security, so we will not delve into the details here. However, given that the server will be a PC, there are plenty of premade implementations of everything we are using on the Rabbit. For reference purposes, we also include the listing of the PC side of the application, written using OpenSSL's implementation of AES and compiled with gcc for Windows, using the Cygwin environment.

Wrap-Up: A Secure Rabbit

We have covered the most important aspects of our Rabbit program in this chapter, but there is a lot more to do to make this sample into a truly secure application. Take a look at the source code listing in Appendix A and see if you can spot any possible security flaws. In any sufficiently large application there are likely a number of flaws that can be exploited in some fashion to varying levels of damage. If you truly want to make it secure, the more eyes you can have look at it, the better. It is also recommended to hire a security professional if you are truly serious about security. The application we have presented here will stop many attackers (especially the lazy ones), but there are numerous aspects of the application that could benefit from additional inspection. For this reason, the application should be viewed as a starting point for a secure application, a foundation to build upon, rather than a secure application in and of itself.

Source Listings

PIC Example Source Code (Main Program)

```
/*******************************************************************
```

Microchip PIC Security Case Study

Timothy Stapko, 2007

This sample program demonstrates the porting of an implementation AES ported from OpenSSL to the PIC using the Microchip MPLAB C18 compiler v3.02 and the MPLAB IDE v7.43.

The security highlighted in this sample is the use of AES with an ordinary TCP/IP network connection to provide a secure mechanism for sending commands to the PIC. The example demonstrates the use of commands to control LEDs on the PICDEM HPC Explorer Board, but the command interface could be adapted to any number of applications.

DISCLAIMER

This sample is for demonstration purposes only. Some code and comments have been adapted from sample programs provided with the MPLAB development environment and for the PICDEM HPC Explorer Board, and are copyrighted by Microchip.

Specific initializations and their associated comments are taken directly from the PIC TCP/IP demo application that comes with the Microchip TCP/IP stack.

The TCP/IP stack for the PIC is also from Microchip: Copyright 2006 Microchip Technology Inc.

This sample application also includes code from the OpenSSL libraries, specificially the AES support libraries, which require the following notices:

Copyright © 1998–2002 The OpenSSL Project. All rights reserved.

This product includes software developed by the OpenSSL Project for use in the OpenSSL Toolkit. (http://www.openssl.org/)

NOTE This sample requires an accompanying sample program intended to be run on a PC using either a Windows environment with the Cygwin environment and GCC, or Linux.

```
*********************************************************************/

    // This define is required to indicate that this file is the primary
    // entry point for the entire application.
    #define THIS_IS_STACK_APPLICATION


    // Header includes for system libraries and TCP stack
    #include <string.h>
    #include "..\Include\Compiler.h"
    #include "..\Include\StackTsk.h"
    #include "..\Include\Tick.h"
    #include "..\Include\MAC.h"
    #include "..\Include\TCP.h"
    #include "..\Include\Helpers.h"
    #include "..\Include\Delay.h"
    #include "..\Include\UART.h"
    #include "..\Include\MPFS.h"
    #include "..\Include\LCDBlocking.h"
    #include "..\Include\PIC.h"


    // AES header include
    // The macro AES_LONG is required to tell the AES libraries to use
    // the "long" type for 32-bit math, since integers in the C18 compiler
    // are natively 16 bits.
    #define AES_LONG
    #include "..\Include\aes.h"


    #if defined(STACK_USE_HTTP_SERVER)
    #include "..\Include\HTTP.h"
    #endif
```

```
// PIC-specific TCP/IP stack initializations

// This is used by other stack elements.

// Main application must define this and initialize it with proper values.

APP_CONFIG AppConfig =

{

    {MY_DEFAULT_IP_ADDR_BYTE1, MY_DEFAULT_IP_ADDR_BYTE2,

     MY_DEFAULT_IP_ADDR_BYTE3, MY_DEFAULT_IP_ADDR_BYTE4},

    {MY_DEFAULT_MAC_BYTE1, MY_DEFAULT_MAC_BYTE2,

     MY_DEFAULT_MAC_BYTE3,

     MY_DEFAULT_MAC_BYTE4, MY_DEFAULT_MAC_BYTE5,

     MY_DEFAULT_MAC_BYTE6},

    {MY_DEFAULT_MASK_BYTE1, MY_DEFAULT_MASK_BYTE2,

     MY_DEFAULT_MASK_BYTE3,

     MY_DEFAULT_MASK_BYTE4},

    {MY_DEFAULT_GATE_BYTE1, MY_DEFAULT_GATE_BYTE2,

     MY_DEFAULT_GATE_BYTE3,

     MY_DEFAULT_GATE_BYTE4}

};


// PIC-specific initializations - this example was written for

// the PIC18F8722 but could easily be adapted to other PIC processors
#if defined(__18CXX)

    #if defined(__18F8722)

        #pragma config OSC=HSPLL, FCMEN=OFF, IESO=OFF, PWRT=OFF, WDT=OFF, LVP=OFF

    #elif defined(__18F87J10) || defined(__18F86J15) ||
defined(__18F86J10) || defined(__18F85J15) || defined(__18F85J10) ||
defined(__18F67J10) || defined(__18F66J15) || defined(__18F66J10) ||
defined(__18F65J15) || defined(__18F65J10)

        #pragma config XINST=OFF, WDTEN=OFF, FOSC2=ON, FOSC=HSPLL

    #elif defined(__18F97J60) || defined(__18F96J65) ||
defined(__18F96J60) || defined(__18F87J60) || defined(__18F86J65) ||
defined(__18F86J60) || defined(__18F67J60) || defined(__18F66J65) ||
defined(__18F66J60)
```

```
            #pragma config XINST=OFF, WDT=OFF, FOSC2=ON, FOSC=HSPLL, ETHLED=ON
        #elif defined(HI_TECH_C)
            __CONFIG(1, HSPLL);
            __CONFIG(2, WDTDIS);
            __CONFIG(3, MCLREN);
            __CONFIG(4, XINSTDIS & LVPDIS);
        #endif
#elif defined(__PIC24F__)
    _CONFIG2(FNOSC_PRIPLL & POSCMOD_HS)        // Primary HS OSC with 4x PLL
    _CONFIG1(JTAGEN_OFF & FWDTEN_OFF)          // JTAG off, watchdog timer off
#elif defined(__dsPIC33F__) || defined(__PIC24H__)
    _FOSCSEL(FNOSC_PRIPLL)               // PLL enabled
    _FOSC(OSCIOFNC_OFF & POSCMD_XT)      // XT Osc
    _FWDT(FWDTEN_OFF)                    // Disable Watchdog timer
    // JTAG should be disabled as well
#elif defined(__dsPIC30F__)
    _FOSC(XT_PLL16)                      // XT Osc + 16X PLL
    _FWDT(WDT_OFF)                       // Disable Watchdog timer
    _FBORPOR(MCLR_EN & PBOR_OFF & PWRT_OFF)
#endif



// PIC-specific initializations for various environments and hardware
// NOTE: Several PICs, including the PIC18F4620 revision A3 have a RETFIE FAST/
MOVFF bug
// The interruptlow keyword is used to work around the bug when using C18
#if defined(HI_TECH_C)
void interrupt HighISR(void)
#else
#pragma interruptlow HighISR
void HighISR(void)
#endif
{
#ifdef __18CXX
```

```
        TickUpdate();
#endif


//#if defined(STACK_USE_SLIP)
//      MACISR();
//#endif
}


// PIC-specific definitions
#if defined(__18CXX) && !defined(HI_TECH_C)
#pragma code highVector=0x08
void HighVector (void)
{
     _asm goto HighISR _endasm
}
#pragma code // Return to default code section
#endif


// Bitmasks for the LED byte
#define CMD_LED1_MASK 0x01
#define CMD_LED2_MASK 0x02
#define CMD_LED3_MASK 0x04
#define CMD_LED4_MASK 0x08
#define CMD_LED5_MASK 0x10
#define CMD_LED6_MASK 0x20
#define CMD_LED7_MASK 0x40


// Bitmasks for the control byte
#define CMD_CHGKEY_MASK        0x01   // Change keys
#define CMD_RESETCTR_MASK      0x02   // Reset global_counter
#define CMD_PRESERVELED_MASK 0x04   // Preserve LED state


// Prototype for command processing function
int ProcessCommand(char * cmd);
```

```
// Global counter to synchronize the PIC with the PC client
unsigned long global_counter;



// The initial AES key. This is unfortunately stored in the
// program memory along with the program itself. It would be
// good to only use this key at startup and then switch
// to a new one stored in a volatile memory (see below).
// Storing keys  on devices and keeping them secret is
// unfortunately a difficult problem to solve without hardware
// support.
// Note that the AES key must be 128 bits (16 bytes)
const rom far char aes_password[] = "PasswordPassword";



// To change the AES password remotely, we need to build a
// state machine to handle the simple protocol we will
// implement below. The reason for the protocol is that
// our command packet is only 16 bytes, which is the same
// size as the AES key - we need multiple messages in
// order to get a full 16-byte key.
// The basic protocol happens in 3 encrypted messages,
// which are received in order:
// 1. A standard command packet (see format in the comment
//    block before ProcessCommand below), with the "change
//    key" bit set in the command byte.
// 2. The very next message received is assumed to be the
//    key. This is saved in the temp_key buffer, but the
//    AES key is left alone.
// 3. A follow-up message is sent that simply contains
//    a checksum of the key. If the checksum checks out,
//    the actual AES key is updated. The reason for this
//    is to verify that we don't change the key to some
//    random 16-byte sequence if the second message is
```

```
//      corrupted or lost.
enum {
    STATE_NORMAL,    // Standard command processing state
    STATE_RECVKEY,   // Next message received is the key
    STATE_CHKSUM     // Next message received is the checksum
};
int current_state;



// Buffer to hold the new candidate key
char temp_key[16];



// For added security, we can send an encrypted message
// back to the PC to make sure everything is working. This
// mechanism can be adapted to send encrypted alerts back to the
// client program on the PC. Note that AES must send things
// in 16-byte blocks, so this message is exactly 16 bytes.
const rom far char aes_message[] = "Command complete";



// AES keys are large and so require additional space. On the PIC, RAM is
// quite limited, so we need to modify the linker scripts to support the
// larger amount of data. To do this, we need to create a data area
// large enough for the key (242 bytes per key for this implementation).
//
// The MAIN_MEM data section was created by updating a couple of lines
// in the 18F8722i.lkr linker script:
//    Comment out these 2 lines:
//       DATABANK    NAME=gpr7    START=0x700    END=0x7FF
//       DATABANK    NAME=gpr8    START=0x800    END=0x8FF
//
//    And Add the following lines - note the difference in the size
//    of gpr8 which is now large enough to hold an AES key or two:
```

```
//      DATABANK    NAME=gpr7       START=0x700         END=0x70F
//      DATABANK    NAME=gpr8       START=0x710         END=0x8FF
//
//    Finally, add the following data section definition at the
//    bottom of the linker script:
//         SECTION    NAME=MAIN_MEM    RAM=gpr8


#pragma udata MAIN_MEM // This pragma enables our new memory area


    // Define 2 keys, one for encrypting outgoing messages and
    // a second for decrypting incoming messages. Note that we
    // could re-use the same key, but it is easier if we have 2,
    // so we don't have to keep clearing and reinitializing one.
    static AES_KEY en_key, dec_key;


#pragma udata            // Return to the existing memory area


//              **************************************************************
// Main application entry point. If using the C30 compiler instead,
// note the different function prototype for main
#ifdef __C30__
int main(void)
#else
void main(void)
#endif
{
    // Program variables.

    // A TCP/IP socket handle
    BYTE my_socket;


    // Buffers for storing messages and keys - note that the AES
    // functions all require passed-in buffers to be in RAM, not
    // in the PIC program memory (more details below when we
```

```
// initialize the AES keys). Note that each buffer is large
// enough to hold an entire AES block (16 bytes each), with
// a little extra.
char input_buf[20], output_buf[20], key_buf[20];


// Various counters, etc...
static int i, data_len;
static TICK t = 0;


// Initialize any application specific hardware.
InitializeBoard();


// Initialize all stack related components.
// Following steps must be performed for all applications using
// PICmicro TCP/IP Stack.
// This initializes the periodic TCP/IP stack tick.
TickInit();


// Following steps must be performed for all applications using
// PICmicro TCP/IP Stack.
// This initializes the PIC filesystem. Note that the filesystem
// is unused in this demo, but the framework is left in since it
// is useful for importing large amounts of data.
MPFSInit();


// Initialize Stack and application related NV variables.
InitAppConfig();


// Initiates board setup process if button is depressed
// on startup
if(BUTTON0_IO == 0)
{
    SetConfig();
}
```

```
    // Initialize the TCP/IP stack
    StackInit();



    // We are using the HTTP server, but not serving pages. Adding
    // an HTML interface would be very simple.
#if defined(STACK_USE_HTTP_SERVER)
    HTTPInit();
#endif



    // Start listening on a TCP/IP socket. The IP address is defined in
    // the TCP/IP stack configuration header (StackTsk.h), and we are
    // using the Telnet port 23 since Telnet is not secure and should
    // really not be used much anymore, but it can easily be changed to
    // something else if you really want Telnet support on your PIC.
    my_socket = TCPListen(23);

    // Initialize variables

    // By default, all of our LEDs start in the OFF state. The HPC
    // Explorer Board has 8 total LEDs, we will utilize LED 0 for
    // a blinking status indicator and the rest are for the example
    LED0_IO = 0;
    LED1_IO = 0;
    LED2_IO = 0;
    LED3_IO = 0;
    LED4_IO = 0;
    LED5_IO = 0;
    LED6_IO = 0;
    LED7_IO = 0;


    // Initialize global_counter (resets the global counter on startup)
    global_counter = 0;
```

```
// Set our current command processing state to the default
current_state = STATE_NORMAL;


// Initialize data buffering
data_len = 0;
memset(input_buf, 0x0, sizeof(input_buf));
memset(output_buf, 0x0, sizeof(output_buf));
memset(key_buf, 0x0, sizeof(key_buf));


// Initialize AES keys. Note that we need to copy the contents
// of the key constant into a RAM buffer stored in the data area.
// The reason for this is that the AES API is setup to access
// the data from RAM, not from program memory ("rom" keyword).
// The API could have been setup to take the ROM constant
// directly, but at some point it will be copied over to RAM
// anyway. Note that we are using 128-bit AES, so the keys are
// 16 bytes each.


// Set up encryption key
memcpypgm2ram(key_buf, aes_password, 16);
AES_set_encrypt_key(key_buf, 128, &en_key);


// Set up decryption key
memcpypgm2ram(key_buf, aes_password, 16);
AES_set_decrypt_key(key_buf, 128, &dec_key);


// PIC-specific comments:
    // Once all items are initialized, go into infinite loop and let
    // stack items execute their tasks.
    // If application needs to perform its own task, it should be
    // done at the end of while loop.
    // Note that this is a "co-operative mult-tasking" mechanism
    // where every task performs its tasks (whether all in one shot
```

```
        // or part of it) and returns so that other tasks can do their
        // job.
        // If a task needs very long time to do its job, it must broken
        // down into smaller pieces so that other tasks can have CPU time.

    while(1)
    {
        // Blink LED0 every second to show we are still running.
        if ( TickGetDiff(TickGet(), t) >= TICK_SECOND/2 )
        {
            t = TickGet();
            LED0_IO ^= 1;
        }


        // Handle TCP/IP tasks
        StackTask();


#if defined(STACK_USE_HTTP_SERVER)
        // Handle HTTP tasks. Note that this example is not currently
        // setup with an HTML interface, but we left the HTTP server
        // framework in so it would be easy to add one.
        HTTPServer();
#endif


        // *** Add other networking handlers here ***
        // Process any I/O on the hardware, other than network traffic
        ProcessIO();


        // The real heart of the application begins here. First, we
        // check to see if our socket has an incoming connection.
        if(TCPIsConnected(my_socket)) {
            // The socket is connected, now see if there is any data
            // to read.
            if(TCPIsGetReady(my_socket)) {
```

```
        // TCPGetArray could be used to grab a block of data
        // at a time from the socket, but the while-loop seems
        // to work better.
        // TCPGetArray(my_socket, input_buf, sizeof(input_buf));
        // Loop until we have filled our input buffer
        while(data_len < sizeof(input_buf) &&
                TCPGet(my_socket, &input_buf[data_len++]))
           ;

        // For safety (and why we added a little extra to
        // the end of the buffer, we terminate the input
        // buffer in case someone decides to use string
        // functions on it.
        input_buf[data_len] = '\0';


        // Clear out the socket, we have our data
        TCPDiscard(my_socket);


        // Now that we have received a command, decrypt it
        // and process it. The decrypted message will be
        // in output_buf when we are done
        AES_decrypt(input_buf, output_buf, &dec_key);


        // Process the incoming command
        ProcessCommand(output_buf);
    }


    // We can send messages back to the PC client from
    // here. Currently, we simply send a message indicating
    // that the command was completed. A possible enhancement
    // to the application would be to have a data collection
    // command that returned the state of the LEDs
    if(TCPIsPutReady(my_socket) && data_len > 0) {
```

```
                    // Encrypt message

                    memcpypgm2ram(input_buf, aes_message, 16);

                    AES_encrypt(input_buf, output_buf, &en_key);

                        i = 0;

                        data_len = 16;

                        while(data_len > 0) {

                            TCPPut(my_socket, output_buf[i]);

                            --data_len;

                            ++i;

                        }

                        TCPFlush(my_socket);

                        data_len = 0;

                        memset(input_buf, 0, sizeof(input_buf));


                    }

                }

            }

        }


// This function processes the incoming command once it has been
// decrypted. First, it checks the counter (bytes 12-15), and then
// the checksum byte (byte 0). Byte 1 contains the info for the
// LED commands (bit location corresponds to the LED to be
// control), byte 2 is for changing the password, bytes 3-7 are
// reserved and should all be 0x0 (they can be used later for other
// commands) and bytes 8-11 are random data used for additional
// protection against replay attacks, and are ignored (they factor
// into the checksum). Finally, bytes 12-15 are a 32-bit counter
// that assures the PIC and PC are synchronized.
//
// The following diagram shows the format of the 16-byte control
// packet sent to the PIC:
//
```

```
// Byte
// 0       | Checksum                                    |
// 1       | LEDs - bits 0-6, bit 8 is ignored
//                    (0 = off, non-zero = on)           |
// 2       | Control byte (other commands)               |
// 3-7     | Reserved (should be 0x0)                    |
// 8-11    | Random bytes (32 bits, ignored)             |
// 12-15   | Counter                                     |
int ProcessCommand(char * cmd) {
    int i;
    unsigned int checksum; // Used for extracting the checksum
    unsigned long counter; // Used for extracting the counter
    int preserve_state;    // flag for the LED preserve command
    int ret_val;           // Return value, -1 is error


    // Return 1 on success
    ret_val = 1;


    switch(current_state) {
    // Default is really an error, but we haven't implemented
    // an elegant way to recover from an improper state error
    // so we fall-through to the basic state.
    default:
    case STATE_NORMAL:


        // Calculate checksum, which adds all bytes (other than the
        // checksum) together and and then truncates to 1 byte
        checksum = 0;
        for(i = 1; i < 16; ++i) {
            checksum += cmd[i];
        }
        // Mask to a single byte
        checksum &= 0xFF;
```

```
if(cmd[0] != checksum) {

    // Error in checksum, return and do nothing

    ret_val = -1;

    break;

}


// Now check counter

counter = ((unsigned long)cmd[12]         ) +

          ((unsigned long)cmd[13] <<    8) +

          ((unsigned long)cmd[14] <<   16) +

          ((unsigned long)cmd[15] <<   24);

if(counter != global_counter) {

    // Counter is not synchronized, return and do nothing

    ret_val -1;

    break;

}


// Increment the global counter

++global_counter;


// Now process the commands, the command packet checks out


// First, check the command byte to see what we are going

// to do. If the bit at the indicated location is set,

// then the command is to be executed.

// Bit position       Command

// ---------------------------------------------------

//      0             Change password

//      1             Reset global_counter

//      2             Only turn LEDs ON (preserve state

//                    of LED if LED bit is not set)

//      3-7           Currently unused


// Change the key - this requires a state machine to grab

// the next 16-byte chunk (containing the new key) and a
```

```
// second chunk with a checksum to ensure that the key
// was transmitted correctly.
if(cmd[2] & CMD_CHGKEY_MASK) {
    // Change the AES key using our state machine. Notice
    // That we don't break out and return, but process the
    // entire command.
    current_state = STATE_RECVKEY;
}

// Reset the counter
if(cmd[2] & CMD_RESETCTR_MASK) {
    global_counter = 0;
}

// If this flag is set, leave LEDs already ON alone,
// and only turn on those LEDs with their bits set in
// the LED byte. This allows the user to turn on a specific
// set of LEDs without having to know the current state of
// the hardware. Another bit could be used to provide the
// opposite effect (turn off specific LEDs and ignore the
// rest, for example).
preserve_state = cmd[2] & CMD_PRESERVELED_MASK;

// Now, control the LEDs
if(preserve_state) {
    // If preserve_state is set, then LEDs retain their
    // current state or are turned on, none are turned off
    LED1_IO = (cmd[1] & CMD_LED1_MASK)? 1 : LED1_IO;
    LED2_IO = (cmd[1] & CMD_LED2_MASK)? 1 : LED2_IO;
    LED3_IO = (cmd[1] & CMD_LED3_MASK)? 1 : LED3_IO;
    LED4_IO = (cmd[1] & CMD_LED4_MASK)? 1 : LED4_IO;
    LED5_IO = (cmd[1] & CMD_LED5_MASK)? 1 : LED5_IO;
    LED6_IO = (cmd[1] & CMD_LED6_MASK)? 1 : LED6_IO;
    LED7_IO = (cmd[1] & CMD_LED7_MASK)? 1 : LED7_IO;
}
```

```
        else {

                // Do not preserve state - interpret the command

                // byte literally (bit set = ON, bit unset = OFF)

                LED1_IO = (cmd[1] & CMD_LED1_MASK)? 1 : 0;

                LED2_IO = (cmd[1] & CMD_LED2_MASK)? 1 : 0;

                LED3_IO = (cmd[1] & CMD_LED3_MASK)? 1 : 0;

                LED4_IO = (cmd[1] & CMD_LED4_MASK)? 1 : 0;

                LED5_IO = (cmd[1] & CMD_LED5_MASK)? 1 : 0;

                LED6_IO = (cmd[1] & CMD_LED6_MASK)? 1 : 0;

                LED7_IO = (cmd[1] & CMD_LED7_MASK)? 1 : 0;

        }

        break;

    case STATE_RECVKEY:  // Next message received is the key

        // Simply copy the message into the temporary buffer

        // and change the state.

        memcpy(temp_key, (void*)cmd, 16);

        current_state = STATE_CHKSUM;

        break;

    case STATE_CHKSUM:   // Next message received is the checksum

        // Calculate the checksum for the stored key and compare

        // to the received checksum

        // NOTE: A single byte checksum may not be sufficient

        // for some applications, and a hash would be even better,

        // but this gets the point across without a lot of extra

        // code.

        checksum = 0;

        for(i = 0; i < 16; ++i) {

            checksum += temp_key[i];

        }



        // Mask checksum byte and compare to received

        if((char)(checksum & 0xFF) == cmd[0]) {
```

```
        // Checksum checks out, update the key

        AES_set_decrypt_key(temp_key, 128, &dec_key);

    }

    else {

        // Error - do nothing and return

        ret_val = -1;

    }


        // Reset our state regardless of the outcome

        current_state = STATE_NORMAL;

        break;

    } // End switch


    // Return to caller

    return ret_val;

}



GCC Source Code (for PIC example)
/************************************************************************

        Microchip PIC Security Case Study - PC client program

        Timothy Stapko, 2007


        This sample program demonstrates the porting of an

        implementation AES ported from OpenSSL to the PIC using the

        Microchip MPLAB C18 compiler v3.02 and the MPLAB IDE v7.43


        The security highlighted in this sample is the use of AES

        with an ordinary TCP/IP network connection to provide a

        secure mechanism for sending commands to the PIC. The

        example demonstrates the use of commands to control LEDs

        on the PICDEM HPC Explorer Board, but the command interface

        could be adapted to any number of applications.
```

```
Files required from OpenSSL:

/crypto/aes/aes_core.c

/crypto/aes/aes_locl.h

/crypto/aes/aes.h


This sample is intended to be compiled with gcc and linked

with aes_core.c from OpenSSL:

    gcc aes_core.c -c

    gcc pic_example.c -c

    gcc pic_example.o aes_core.o


This example was compiled on a Windows PC using gcc in the

Cygwin environment (www.cygwin.com), but should compile

under Linux with little or no modification.


***DISCLAIMER***

This sample is for demonstration purposes only.


This sample application includes code from the OpenSSL

libraries, specificially the AES support libraries, which

require the following notices:


    Copyright (c) 1998-2002 The OpenSSL Project.  All rights

    reserved.


    This product includes software developed by the OpenSSL

    Project for use in the OpenSSL Toolkit.

    (http://www.openssl.org/)

 ********************************************************************/
// Standard includes
#include <stdio.h>

#include <sys/socket.h>

#include <arpa/inet.h>

#include <stdlib.h>
```

```
#include <string.h>

#include <unistd.h>


// Include AES support from OpenSSL

#include "aes.h"


// Size of buffer for sending command messages

#define BUFSIZE 20


// Bitmasks for the LED byte

#define CMD_LED1_MASK 0x01

#define CMD_LED2_MASK 0x02

#define CMD_LED3_MASK 0x04

#define CMD_LED4_MASK 0x08

#define CMD_LED5_MASK 0x10

#define CMD_LED6_MASK 0x20

#define CMD_LED7_MASK 0x40


// Bitmasks for the control byte

#define CMD_CHGKEY_MASK        0x01   // Change keys

#define CMD_RESETCTR_MASK      0x02   // Reset global_counter

#define CMD_PRESERVELED_MASK 0x04   // Preserve LED state


// Byte locations for command options

#define CHKSUM_BYTE 0

#define LED_BYTE      1

#define CMD_BYTE      2

#define CTRLOW_BYTE 12


// Keep a global array containing our current command,

// which will be constructed by the user before sending.

char current_command[16];


// Temporary key storage for key changeover

char temp_key[16];
```

```
    // 32-bit integer counter for integrity checking in
    // command packet
    int global_counter;


    // Prototypes
    void ProcessCommand(void);
    void ProcessKeyChange(void);
    void ProcessLeds(void);


    // Error handler
    void ErrorExit(char *msg)
    {
        perror(msg);
        exit(1);
    }


    // The main program.
    // The program establishes a connection to the target (address provided
    // from the command line) and then proceeds to enter a loop and ask the
    // user for input. The program then constructs a command packet, encrypts
    // it using the current AES key, and sends it to the target device.
    // Note that this program is not currently set up to receive messages
    // from the target device, but easily could be modified to do so.
    int main(int argc, char *argv[])
    {
        int sock, i;                        // Socket handle
        int cmd;                            // Command selected
        int key_change;                     // Flag for key changeover operation
        struct sockaddr_in target_addr;     // Target IP address structure
        unsigned short target_port;         // Target TCP port
        char *target_ip;                    // Target IP address (string)
        char send_buf[BUFSIZE];             // Buffer for sending data
```

```
unsigned int echoStringLen;      // Length of string to echo

int data_len;                    // Bytes read in single recv()


AES_KEY en_key, dec_key;         // AES key structures

char output_buffer[128];         // buffer for printing

char print_buffer[128];


// Test for correct arguments

if ((argc < 2) || (argc > 3))

{

    fprintf(stderr, "Usage: %s <Target IP> <Target Port>\n",

            argv[0]);

    exit(1);

}


// First argument: target IP address (dotted quad format)

target_ip = argv[1];


// Second argument: TCP port

target_port = atoi(argv[2]);


// Open the TCP/IP socket to the target device - start by getting

// a socket handle

sock = socket(PF_INET, SOCK_STREAM, IPPROTO_TCP);

if(sock < 0) {

    ErrorExit("Could not get socket handle.");

}


// Fill in the address data structure

memset(&target_addr, 0, sizeof(target_addr));    // Clear out structure

target_addr.sin_family      = AF_INET;           // Internet address

target_addr.sin_addr.s_addr = inet_addr(target_ip); // Target IP address

target_addr.sin_port        = htons(target_port);    // Target port


// Open the socket to the target device

if (connect(sock, (struct sockaddr *) &target_addr,
```

```
            sizeof(target_addr)) < 0)
{
    ErrorExit("Socket establishment failed.");
}


// Now that we have a socket, setup the AES keys
AES_set_encrypt_key("PasswordPassword", 128, &en_key);
AES_set_decrypt_key("PasswordPassword", 128, &dec_key);


// Clear out our current command to initialize
memset(current_command, 0x0, sizeof(current_command));


// Setup our global counter
global_counter = 0;


// Clear the key change flag
key_change = 0;


// Command loop
while(1) {
    printf("Pick a command:\n");
    printf("  1: Select LEDs to turn on or off\n");
    printf("  2: Set 'preserve LED state' flag\n");
    printf("  3: Reset global counter\n");
    printf("  4: Change AES key\n");
    printf("  5: Reset command (start over)\n");
    printf("  6: Print current command\n");
    printf("  7: Send current command to target\n");
    printf("  8: Quit\n>");
    scanf("%d", &cmd);

    switch(cmd) {
    case 1:
    ProcessLeds();
        break;
```

```
case 2:

    // Preserve LED states (don't turn off LEDs,
// just turn them on)
current_command[CMD_BYTE] |= CMD_PRESERVELED_MASK;

    break;
case 3:
// Reset the global counter

global_counter = 0;

current_command[CMD_BYTE] |= CMD_RESETCTR_MASK;

    break;
case 4:
// Process a key changeover
current_command[CMD_BYTE] |= CMD_CHGKEY_MASK;

ProcessKeyChange();

key_change = 1;

    break;
case 5:
    memset(current_command, 0x0, sizeof(current_command));

    break;
case 6:
printf("-----------------------------------------------------------\n");

for(i = 1; i < 8; ++i) {

    printf("LED %d: %s\n", i,

        (current_command[LED_BYTE] & (0x1 << (i-1)))? "On" : "Off");

}

    printf("Preserve LED state: %s \n",

    (current_command[CMD_BYTE] & CMD_PRESERVELED_MASK)?

     "Set" : "Not set");

printf("Reset counter: %s \n",

   (current_command[CMD_BYTE] & CMD_RESETCTR_MASK)?

   "Set" : "Not set");

printf("Change AES key: %s \n",
```

```
                    (current_command[CMD_BYTE] & CMD_CHGKEY_MASK)?
        "Set" : "Not set");
    printf("--------------------------------------------------------\n\n");
     break;
    case 7:
        printf("Sending command...\n");
        ProcessCommand();

        printf("Command: ");
        for(i = 0; i < 16; ++i) {
            printf("%02X ", current_command[i] & 0xff);
        }
        printf("\n");

        // Encrypt the outgoing message
        AES_encrypt(current_command, output_buffer, &en_key);
        if (send(sock, output_buffer, 16, 0) != 16) {
            ErrorExit("Error: send() sent a different
            number of bytes than expected");
        }

        if(key_change) {
            char response[2];
            // Follow the key-change protocol:
            // 1) Send the command (done already)
            // 2) Send the new key encrypted using the old one
            AES_encrypt(temp_key, output_buffer, &en_key);
            if (send(sock, output_buffer, 16, 0) != 16) {
             ErrorExit("Error: send() sent a different number of bytes than
              expected");
            }

            // 3) Calculate the checksum of the key for the third message
            memset(output_buffer, 0x0, sizeof(output_buffer));
```

```
        for(i = 0; i < 16; i++) {

            output_buffer[0] += temp_key[i];

        }

        // Encrypt the checksum message and send

        AES_encrypt(output_buffer, output_buffer, &en_key);

        printf("About to change AES key, continue? (y or n):

        ");

        scanf("%s", &response);

        if(response[0] == 'y') {

            if (send(sock, output_buffer, 16, 0) != 16) {

                    ErrorExit("Error: send() sent a different

                    number of bytes than expected");

            }



            // Switch over to the new key

            AES_set_encrypt_key(temp_key, 128, &en_key);

        }

        key_change = 0;

    }



    // Wait for reply from target

    bytes_rcvd = i = 0;

    printf("Waiting for response...\n");

    while (bytes_rcvd < 16)

    {

        printf("\nBegin receiving:\n");

        i = recv(sock, recv_buf, BUFSIZE - 1, 0);

// Receive from target

        if (i <= 0) {

        ErrorExit("recv() failed or connection closed prematurely");

        }
```

```
                    bytes_rcvd += i;    // Total bytes received

            }

       // Decrypt the incoming message
       AES_decrypt(recv_buf, output_buffer, &dec_key);


     // Print the string from the target
     output_buffer[16] = 0x0;
     printf("Recevied from target: %s\n", output_buffer)

        break;
        case 8:
            close(sock);
        exit(0);
        default:
        printf("Unrecognized command\n");
            break;

        }

    }
    close(sock);
    exit(0);
}


// This function takes the current command constructed by the user
// and formats it for the command protocol to be sent over to
// the PIC. This primarily consists of adding the counter to the
// current command buffer and calculating the final checksum.
// The command will be encrypted and sent to the PIC back in the
// primary command loop in main
void ProcessCommand(void) {
    int checksum, i;


    // Put the global counter into the current command buffer
    current_command[CTRLOW_BYTE] =      (char)((global_counter      ) & 0xFF);
    current_command[CTRLOW_BYTE + 1] = (char)((global_counter >>  8)  &
```

```
OxFF);

    current_command[CTRLOW_BYTE + 2] = (char)((global_counter >> 16)  &

OxFF);

    current_command[CTRLOW_BYTE + 3] = (char)((global_counter >> 24)  &

OxFF);


    // Calculate the checksum byte and add it to the current command buffer
    checksum = 0;
    for(i = 1; i < 16; ++i) {
        checksum += current_command[i];
    }
    current_command[CHKSUM_BYTE] = (char)(checksum & OxFF);


    // Finally, increment the global counter
    ++global_counter;
}


// This function simply requests a new AES key from the user, which
// is entered as ASCII text. This isn't the greatest way to create
// a key, but for the purposes of this example, it works just fine.
void ProcessKeyChange(void) {
printf("Enter new AES key (must be exactly 16 characters):\n");
    scanf("%16s", temp_key);
}


// Process the LEDs to turn on and off for the example. This function
// prompts the user to enter which LED to change, followed by the
// value to change the LED to (1=ON, 0=OFF). We don't receive anything
// back from the PIC, but an improvement could be to read the LED state
// on the PIC and report it back to this program for more interactivity.
void ProcessLeds(void) {
    int led, i;
    char new_byte;
```

```c
        // Keep processing requests until the user is done
while(1) {
    printf("Enter which LED you want to change, 1-7\n");
    printf("Enter 0 to return\n>");
    scanf("%d", &led);
    if(led == 0x0) {
        break;
    }

    // If the number is appropriate, prompt the user for the value
    if(led >= 1 && led <= 7) {
        // Subtract 1 to get the right bit position
        --led;
        while(1) {
            printf("Enter 1 for ON, 0 for OFF >");
            scanf("%d", &i);
            if(i != 0 && i != 1) {
                printf("You must enter 0 or 1\n");
            }
            else {
                // Clear the bit
                new_byte = current_command[LED_BYTE] & ~((char)0x1 << led);

                // Set the bit
                current_command[LED_BYTE] = new_byte | (char)(i << led);

                break;
            }
        }
    }
    else {
        printf("Please enter an LED number between 1 and 7.\n");
    }
}
}
```

Rabbit Source Code (Chapter 15 case study)

```
/*******************************************************************************

        Rabbit Security Case Study

        Vending Machine Example

        Timothy Stapko, 2007


        This sample program demonstrates some of the security features

        provided by Rabbit Semiconductor for their Rabbit 4000

        microprocessor products. This sample was specifically designed to

        run on an RCM4010 Rabbit Core Module with Ethernet.


        The security features highlighted in this sample include the

        Secure Sockets Layer (SSL) add-on software module, the

        Advanced Encryption Standard (AES) add-on software module,

        and the tamper-detection feature of the Rabbit 4000 CPU.


        ***DISCLAIMER***

        This sample is for demonstration purposes only. Some code and

        comments have been adapted from sample programs provided with

        the Dynamic C development environment, and are copyrighted by

        Rabbit Semiconductor and Digi International.


        ***NOTE*** This sample will NOT compile without first creating a

        certificate with the name and path specified in the #ximport line

        ("#ximport 'cert\mycert.dcc' SSL_CERTIFICATE") below. See the SSL

        Walkthrough, Section 4.1 in the Dynamic C SSL User's Manual for

        information on creating certificates for SSL programs.


*******************************************************************************/
#class auto

// Pick the predefined TCP/IP configuration for this sample.  See
// LIB\TCPIP\TCP_CONFIG.LIB for instructions on how to set the
// configuration.
#define TCPCONFIG 1
```

```
// ----- Web server configuration -----


// only one socket and server are needed for a reserved port,
// but to use HTTP and HTTPS on the same device, 2 servers
// are needed, one for HTTP and one for HTTPS
#define HTTP_MAXSERVERS 2
#define MAX_TCP_SOCKET_BUFFERS 2


// By default, digest authentication is turned off.  Note that you
// can set USE_HTTP_BASIC_AUTHENTICATION to 0 to remove the code for
// basic authentication at compile time.
#define USE_HTTP_DIGEST_AUTHENTICATION   1


// This macro determines the number of HTTP servers that will use
// SSL (HTTPS servers). In this case, we have 2 total servers, and
// this defines one of them to be HTTPS
#define HTTP_SSL_SOCKETS 1


// Size of the buffer that will be allocated to do error
// processing during form parsing.  This buffer is allocated
// in root RAM.  Note that it must be large enough to hold
// the name, value, and 4 more bytes for each variable in the
// form.
// This parameter must be defined in order to use the form
// generation and parsing functionality.
#define FORM_ERROR_BUF 256


// Define this if you do not need to use the http_flashspec array.
// Since this program uses only the ZSERVER.LIB functionality,
// then it does not need the http_flashspec array.
#define HTTP_NO_FLASHSPEC


// This program does not need to do any DNS lookups, so it
// disables them in order to save memory.  Note that resolve()
```

```
// can still handle IP addresses with this option defined, but
// it can not handle names.
#define DISABLE_DNS

// SSL Stuff
// This macro tells the HTTP library to use SSL
#define USE_HTTP_SSL

// Import the SSL certificate to be stored with the program
#ximport "mycert.dcc" SSL_CERTIFICATE

/*******************************
 * End of configuration section *
 *******************************/

#memmap xmem
#use "dcrtcp.lib"      // TCP/IP library
#use "http.lib"        // HTTP support
#use "aes_crypt.lib"   // AES library
#use "RCM40xx.LIB"     // Hardware support

// the default mime type for '/' must be first
const HttpType http_types[] =
{
    { ".html", "text/html", NULL}
};

// Macro definitions for server address information. These could be changed
// to variables that could be modified using the HTTP interface.
#define SERVER_PORT    8443
#define SERVER_IP_ADDR "10.10.6.105"

// Used for tamper-detect RAM copies
#define AES_IV_OFFS 16

// Status variables for the applicaton should be declared here
// ...
```

```
// The AES passphrase buffer. This array is associated with the HTML
// textbox in which the administrator will enter the new AES passphrase
static char aes_passphrase[32];


// String for initialization vector to be entered by the user
static char aes_iv_phrase[32];


// The actual AES key used for encrypting outgoing traffic. The key is
// an MD5 hash of the passphrase - the size of an MD5 hash is always
// 16 bytes
static char aes_key[16];


// AES initialization vector. This is vital as we are using AES as a stream
// cipher.
static char aes_init_vector[16];


// AES stream state structure (AES-CFB)
AESstreamState aes_state;


// MD5 state variable for hashing the AES passphrase
static md5_state_t aes_md5_state;


// The web interface password. This array is associated with the HTML
// textbox in which the administrator can enter a new password for
// the web interface. Note that the administrator user name does not change.
static char web_password[32];
static char repeat_password[32];


// Keep a global array containing our current message,
// which will be constructed by the application before sending.
static char current_message[100];


// 32-bit integer counter for integrity checking in
// command packet
static int global_counter;
```

```
// Check the web password textbox by comparing its value to the repeat
// password textbox value
int checkRepeatPassword(char * newval) {
    auto char * var;
    auto char * value;
    auto int result;


    // Get the value of the web_password textbox
    var = http_finderrbuf("web_password");
    if (var != NULL) {
        // Get the value of the the textbox and compare to the new value
        // for this textbox
        http_nextfverr(var, NULL, &value, NULL, NULL);
        result = strncmp(newval, value, sizeof(web_password));
    }
    else {
        return -1;
    }


    // Check the result of checking the new password
    if(result == 0x0) {
        // Repeat and new value are the same
        return 0;
    }
    // Repeat password does not match primary password
    return -1;
}



// This function generates a message to be sent to the server.
// The format for our data packets:
// Byte   0          1         5       6          10    10 + n
//        | Checksum | Counter | Size | Random | Data |
```

```c
#define PKT_CHKSUM_BYTE 0

#define PKT_CTRLOW_BYTE 1

#define PKT_SIZE_BYTE   5

#define PKT_RANDOM_BYTE 6

#define PKT_HDR_SIZE    10


int GenerateMessage(char * msg, int msg_sz, char * data, int data_sz) {
    auto int checksum, i;
    auto float rand_value;

    #GLOBAL_INIT {
        // Dynamic C initialization for static variables (executed at
        // program startup).
        global_counter = 0;
    }


    if(msg_sz < data_sz + PKT_HDR_SIZE) {
        printf("Error: Message buffer too small.\n");
        return -1;
    }


    // Put the global counter into the current command buffer
    msg[PKT_CTRLOW_BYTE] =      (char)((global_counter       ) & 0xFF);
    msg[PKT_CTRLOW_BYTE + 1] = (char)((global_counter >>  8) & 0xFF);
    msg[PKT_CTRLOW_BYTE + 2] = (char)((global_counter >> 16) & 0xFF);
    msg[PKT_CTRLOW_BYTE + 3] = (char)((global_counter >> 24) & 0xFF);


    // The size of our data is put into the packet (max of 256
    // characters of data)
    msg[PKT_SIZE_BYTE] = data_sz;


    // Put the random data into our packet
    // We use the "rand()" function here, which is probably not
    // cryptographically secure. Using a better random function here
    // would improve the security of this mechanism
```

```
    rand_value = rand();

    msg[PKT_RANDOM_BYTE] =        (char)(((long)rand_value       ) & 0xFF);

    msg[PKT_RANDOM_BYTE + 1] = (char)(((long)rand_value >>  8) & 0xFF);

    msg[PKT_RANDOM_BYTE + 2] = (char)(((long)rand_value >> 16) & 0xFF);

    msg[PKT_RANDOM_BYTE + 3] = (char)(((long)rand_value >> 24) & 0xFF);


    // Put the data into the message

    for(i = 0; i < data_sz; i++) {

        msg[i + PKT_HDR_SIZE] = data[i];

    }


    // Calculate the checksum byte and add it to the packet

    // Don't include the checksum byte itself

    // We could use a hash algorithm here, but it would probably be overkill

    // for this application

    checksum = 0;

    for(i = 1; i < data_sz + PKT_HDR_SIZE; ++i) {

        checksum += msg[i];

    }

    msg[PKT_CHKSUM_BYTE] = (char)(checksum & 0xFF);


    // Finally, increment the global counter

    ++global_counter;


    // Return the size of the message

    return data_sz + PKT_HDR_SIZE;

}


#define DS2_BIT 2

#define DS3_BIT 3

#define S2_BIT  4

#define S3_BIT  5

static int switch_val;


// Application-specific processing

int processApplication() {
```

```
    // Variable declarations
    int _new_status;

    // If status changes, the variable _new_status == 1
    _new_status = 0;

    return _new_status;
}


// This function takes a message and sends it to the defined server.
void SendMessage(char * msg, int msg_sz) {
    // Server socket variables
    auto char   buffer[100];
    auto int    bytes_read;
    auto tcp_Socket socket;
    auto long serverIP;

    serverIP = resolve(SERVER_IP_ADDR);
    if(serverIP == 0L) {
        printf("ERROR: Cannot resolve \"%s\" into an IP address\n",
               SERVER_IP_ADDR);
        return;
    }


    // Open the connection
    tcp_open(&socket, 0, serverIP, SERVER_PORT, NULL);

    // Wait for the socket to be established
    printf("Waiting for connection...\n");
    while(!sock_established(&socket) && sock_bytesready(&socket)==-1) {
        tcp_tick(NULL);
    }


    printf("Connection established, sending status message...\n");
```

```
    // Send the message to the server
    sock_write(&socket, msg, msg_sz);

    // Send the message, wait for verification from server
    do {
        bytes_read = sock_fastread(&socket,buffer,sizeof(buffer)-1);

        if(bytes_read>0) {
            buffer[bytes_read] = '\0';
        }
    } while(tcp_tick(&socket));

    // Close our server socket
    sock_abort(&socket);
    printf("\nConnection closed...\n");
}


// This function simply encrypts a message in a buffer in place (wrapper
// for the AES API function). This application is designed for a single
// AES state, so this function helps to enforce that idea.
void EncryptMessage(char * msg, int msg_sz) {
    // Encrypt data in place
    AESencryptStream(&aes_state, msg, msg_sz);
}


// ---------- M A I N ----------
void main(void)
{
    // Declare the FormVar array to hold form variable information
    auto FormVar myform[5];
    auto int admin_user;
    auto int result;
    auto int var;
    auto int form;
    auto int i;
```

```
auto int status_changed; // Flag to indicate when messages should be sent
auto int msg_sz;

// Initialize hardware
brdInit();
switch_val = 0;              //initialize switch to false value

// Initialize variables
memset(&aes_passphrase, 0, sizeof(aes_passphrase));
memset(&aes_key, 0, sizeof(aes_key));
memset(&aes_init_vector, 0, sizeof(aes_init_vector));
memset(&aes_iv_phrase, 0, sizeof(aes_iv_phrase));
memset(&aes_md5_state, 0, sizeof(aes_md5_state));
memset(&web_password, 0, sizeof(web_password));
memset(repeat_password, 0, sizeof(repeat_password));

// HTTP_DIGEST_AUTH is the default authentication type when
// digest authentication has been enabled, so this line is not
// strictly necessary.  The other possible values are
// HTTP_BASIC_AUTH and HTTP_NO_AUTH.
http_setauthentication(HTTP_DIGEST_AUTH);
printf("Using digest authentication\n");

// Create the administrator user, with default password
admin_user = sauth_adduser("admin", "password", SERVER_HTTPS);

// Add the form (array of variables)
// Note the use of SERVER_HTTPS instead of SERVER_HTTP. This means that
// the form declared here is only accessible from an SSL-secured HTTPS
// socket. If you want this form to be accessible from both HTTPS and
// HTTP clients, then the final parameter would be
// SERVER_HTTPS | SERVER_HTTP
form = sspec_addform("myform.html", myform, 5, SERVER_HTTPS);

// Add the authorized administrator to the users that can access the page
sspec_adduser(form, admin_user);
```

```
// Add the form to the "Admin" realm so that it can only be accessed by
// administrators
sspec_setrealm(form, "Admin");


// Set the title of the form
sspec_setformtitle(form, "Vending Machine Administration Interface");


// Add the AES passphrase variable, and set it up with the form
var = sspec_addvariable("aes_passphrase", &aes_passphrase, PTR16, "%s",
                                      SERVER_HTTPS);

var = sspec_addfv(form, var);

sspec_setfvname(form, var, "AES Key");

sspec_setfvdesc(form, var, "Enter AES passphrase (32 characters max)");

sspec_setfvlen(form, var, 32);

sspec_setfventrytype(form, var, HTML_FORM_TEXT);


// Init vector entry box
var = sspec_addvariable("aes_iv_phrase", &aes_iv_phrase, PTR16, "%s", SERVER_HTTPS);

var = sspec_addfv(form, var);

sspec_setfvname(form, var, "AES Initialization vector");

sspec_setfvdesc(form, var, "Enter initialization vector (16 characters)");

sspec_setfvlen(form, var, 32);

sspec_setfventrytype(form, var, HTML_FORM_TEXT);


// Change SSL login password for HTTP interface
var = sspec_addvariable("web_password", &web_password, PTR16, "%s",
                                      SERVER_HTTPS);

var = sspec_addfv(form, var);

sspec_setfvname(form, var, "Change password");

sspec_setfvdesc(form, var, "Enter new password (32 characters max)");

sspec_setfvlen(form, var, 32);

sspec_setfventrytype(form, var, HTML_FORM_TEXT);


var = sspec_addvariable("repeat_password", &repeat_password, PTR16, "%s",
                                      SERVER_HTTPS);
```

```
    var = sspec_addfv(form, var);

    sspec_setfvname(form, var, "Repeat Password");

    sspec_setfvdesc(form, var, "Repeat new password");

    sspec_setfvlen(form, var, 32);

    sspec_setfventrytype(form, var, HTML_FORM_TEXT);

    sspec_setfvcheck(form, var, checkRepeatPassword);


    // Create aliases for this form.  This allows the form to be accessed from
    // other locations.
    sspec_aliasspec(form, "index.html");

    sspec_aliasspec(form, "/");


    sock_init();

    http_init();

    tcp_reserveport(80);


    // We need to reserve the HTTPS port as well, which is TCP port 443
    tcp_reserveport(443);


    // Clear our passphrase buffer, key, and initialization vector
    memset(&aes_passphrase, 0, sizeof(aes_passphrase));

    memset(&aes_key, 0, sizeof(aes_key));

    memset(&aes_init_vector, 0, sizeof(aes_init_vector));


    // Clear flags and other information
    status_changed = 0;


    // Primary program loop - all processing is done in this loop
    while (1) {

        http_handler();


        // Check for new password
        if (web_password[0] != 0 || repeat_password[0] != 0) {

        result = strncmp(web_password, repeat_password, sizeof(repeat_password));

            if(result == 0x0) {
```

```
        // Set the new password for the admin
        printf("Password changed\n");
        sauth_setpassword(admin_user, web_password);

        // Clear out the textbox variables for security purposes
    memset(web_password, 0, sizeof(web_password));
    memset(repeat_password, 0, sizeof(repeat_password));
    }
}


// Check to see if the user has entered a new passphrase to generate
// a new AES key for outgoing information
if(aes_passphrase[0] != 0x0) {
// Initialize the MD5 hash
md5_init(&aes_md5_state);

// Add the AES passphrase from the textbox to the MD5 hash
md5_append(&aes_md5_state, aes_passphrase, strlen(aes_passphrase));
    // Finish the hash and put the result in the AES key buffer
    md5_finish(&aes_md5_state, aes_key);

    // For demonstration purposes, print out the new AES key
    printf("AES Passphrase:%s\n", aes_passphrase);
    for(i = 0; i < sizeof(aes_key); ++i) {
        printf("AES Key[%d]:%d\n", i, aes_key[i]);
    }
    printf("\n");

    // Extract the AES IV temporarily from tamper-detect RAM
    vram2root(aes_init_vector, AES_IV_OFFS, sizeof(aes_init_vector));

    // Setup the AES state with the new key and existing IV
    AESinitStream(&aes_state, aes_key, aes_init_vector);

    // Put key in tamper-detect RAM for safe-keeping
    root2vram(aes_key, 0, sizeof(aes_key));
```

```
        // Clear out all temporary data buffers associated with the AES
        // passphrase and key generation. The key is now stored only in
        // the tamper-detect RAM register
        memset(&aes_passphrase, 0, sizeof(aes_passphrase));
        memset(&aes_md5_state, 0, sizeof(aes_md5_state));
        memset(&aes_key, 0, sizeof(aes_key));
    }

    // Check for a new AES initialization vector
    if(aes_iv_phrase[0] != 0x0) {
        // Initialize the new initialization vector
        strncpy(aes_init_vector, aes_iv_phrase, sizeof(aes_init_vector));

        // For demonstration purposes, print out the data
        printf("AES Initialization Vector phrase:%s\n", aes_iv_phrase);
        for(i = 0; i < sizeof(aes_init_vector); ++i) {
            printf("AES IV[%d]:%d\n", i, aes_init_vector[i]);
        }
        printf("\n");

        // Extract the AES key temporarily from tamper-detect RAM
        vram2root(aes_key, 0, sizeof(aes_key));

        // Setup the AES state with the new IV and existing key
        AESinitStream(&aes_state, aes_key, aes_init_vector);

        // Put the IV into tamper-detect RAM for safe-keeping
        root2vram(aes_init_vector, AES_IV_OFFS, sizeof(aes_init_vector));

        // Clear out our buffers
        memset(&aes_init_vector, 0, sizeof(aes_init_vector));
        memset(&aes_iv_phrase, 0, sizeof(aes_iv_phrase));
    }

    // Application-specific code goes in this function. If there is status to
    // be sent to the centralized server, the flag is set to 1.
```

```
status_changed = processApplication();

// The keywords "costate", "abort" and "waitfor" are specific to Dynamic C
// and are used for building simple state machines. We are using them here
// to see if a switch on the development board is pressed, and turning
// on an LED to indicate the LED was pressed.
// This mechanism is used for testing purposes, but it illustrates the
// use of the hardware with secure communications.
costate
{
    if (BitRdPortI(PBDR, S2_BIT)) {   // Wait for switch S2 press
        abort;
    }
    // Switch press detected if get to here
    waitfor(DelayMs(50));
    // Turn on LED while switch is held down
    switch_val = 1;

    if (BitRdPortI(PBDR, S2_BIT))
    {
        // Wait for switch release and then turn LED off, reset switch

        // and notify the state machine that there is a message to send
        switch_val = !switch_val;
        status_changed = 1;
        abort;
    }
}

costate {
    if(switch_val) {
        BitWrPortI(PBDR, &PBDRShadow, 0, DS2_BIT);
    }
    else {
```

```
                // Turn on LED to indicate switch press
                BitWrPortI(PBDR, &PBDRShadow, 1, DS2_BIT);
            }
        }


        // Send information using AES encryption to the centralized server
        // program.
        if(status_changed) {
            // Create the new message. This currently just uses a string literal
            // but it can easily be modified to send any data you want.
            msg_sz = GenerateMessage(current_message, sizeof(current_message),
                        "Status Message", 14);

            // If GenerateMessage produces an error (return value < 0), then
            // don't do anything and leave flag alone so we can try again
            if(msg_sz > 0) {
                // Encrypt our new message in place using the current key
                EncryptMessage(current_message, msg_sz);

                // Send the message to the target server
                SendMessage(current_message, msg_sz);

                // Clear out our message after it has been sent, update status flag
                memset(current_message, 0x0, sizeof(current_message));
                status_changed = 0;
            }
        }
    }
}
// END PROGRAM


/*********************************************************************
        Rabbit Security Case Study - PC client program
        Timothy Stapko, 2007
        Vending Machine Example
```

This sample program demonstrates some of the security features provided by Rabbit Semiconductor for their Rabbit 4000 microprocessor products.

This sample is a simple PC-based server that accepts incoming messages from the Rabbit. The logic is simple to show what is being done, but it could easily be adapted to be an active server.

DISCLAIMER
This sample is for demonstration purposes only. Some of the code was written for use with some code from OpenSSL:

Files required from OpenSSL:
/crypto/aes/aes_core.c
/crypto/aes/aes_locl.h
/crypto/aes/aes_cfb.c
/crypto/aes/aes.h

This sample is intended to be compiled with gcc and linked with aes_core.c from OpenSSL:

    gcc aes_core.c -c
    gcc aes_cfb.c -c
    gcc rabbit_example.c -c
    gcc rabbit_example.o aes_core.o aes_cfb.o

This example was compiled on a Windows PC using gcc in the Cygwin environment (www.cygwin.com), but should compile under Linux with little or no modification.

DISCLAIMER
This sample is for demonstration purposes only.

This sample application includes code from the OpenSSL libraries, specificially the AES support libraries, which require the following notices:

```
        ****************************************************************/

// Standard includes
#include <stdio.h>
#include <sys/socket.h>
#include <arpa/inet.h>
#include <stdlib.h>
#include <string.h>
#include <unistd.h>

// Include AES support from OpenSSL
#include "aes.h"

// 32-bit integer counter for integrity checking in messages
int global_counter;

// Error handler
void ErrorExit(char *msg)
{
    perror(msg);
    exit(1);
}

// The format for our data packets:
// Byte    0           1        5       6        10    10 + n
//         | Checksum | Counter | Size | Random | Data |
#define PKT_CHKSUM_BYTE 0
#define PKT_CTRLOW_BYTE 1
```

```
#define PKT_SIZE_BYTE    5

#define PKT_RANDOM_BYTE 6

#define PKT_HDR_SIZE     10


// Index and feedback buffer for AES CFB mode

int cfb_index = 0;

char cfb_buf[16];


// Function prototypes

void GetAESKey(AES_KEY *key);

int ParseMessage(char * msg, int msg_sz, char * data, int data_sz);

void AES_CFB_decrypt(AES_KEY *key, char *data, int size);


// The main program.

// Program sets up a server and listens for incoming requests.

// Upon receiving a request and establishing a link, the program

// recevies, decrypts, and processes an incoming message.

// This is meant as a demonstration for debugging purposes, the

// interesting code is on the Rabbit.

int main(int argc, char *argv[])

{

    int sock, i;                     // Socket handle

    int err;                         // Variable for error handling

    int listen_handle;               // Socket listen handle

    struct sockaddr_in server_sock;  // Server structure

    struct sockaddr_in client_sock;  // Client structure

    size_t client_len;               // Length of client structure


    unsigned short target_port;      // Target TCP port

    char send_buf[BUFSIZE];          // Buffer for sending data

    char recv_buf[BUFSIZE];          // Buffer for receiving data

    unsigned int bytes_rcvd;         // Number of bytes from target

    int data_len;                    // Bytes read in single recv()
```

```c
    AES_KEY dec_key;                    // AES key structure
    char output_buffer[128];            // buffer for printing
    char print_buffer[128];

    // Test for correct arguments
    if ((argc < 1) || (argc > 2))
    {
        fprintf(stderr, "Usage: %s <Target Port>\n",
                argv[0]);
        exit(1);
    }


    // First argument: TCP port
    target_port = atoi(argv[1]);

    // Prompt user for AES key and Initialization vector
    GetAESKey(&dec_key);

    // Setup socket
    listen_handle = socket (AF_INET, SOCK_STREAM, 0);
    if(err == -1) {
        printf("Error in call to socket");
    }


    // Socket data structure
    memset (&server_sock, '\0', sizeof(server_sock));
    server_sock.sin_family      = AF_INET;
    server_sock.sin_addr.s_addr = INADDR_ANY;
    server_sock.sin_port        = htons (target_port);

    // Setup the port and listen for incoming messages
    err = bind(listen_handle, (struct sockaddr*) &server_sock,
          sizeof (server_sock));
    if(err == -1) {
        printf("Error in call to bind");
    }
```

```
err = listen (listen_handle, 5);
if(err == -1) {
    printf("Error in call to listen");
}


// Accept incoming connection
client_len = sizeof(client_sock);
sock = accept (listen_handle, (struct sockaddr*) &client_sock, &client_len);
if(sock == -1) {
    printf("Error in call to accept");
}
close (listen_handle);


// Open the TCP/IP socket to the target device - start by getting
// a socket handle
sock = socket(PF_INET, SOCK_STREAM, IPPROTO_TCP);
if(sock < 0) {
    ErrorExit("Could not get socket handle.");
}


// Setup our global counter
global_counter = 0;


printf("Waiting for response...\n");
while (bytes_rcvd < 16)
{
    printf("\nBegin receiving:\n");
    i = recv(sock, recv_buf, BUFSIZE - 1, 0);
    // Receive from target
    if (i <= 0) {
        ErrorExit("recv() failed or connection closed prematurely");
    }
    bytes_rcvd += i;    // Total bytes received
}
```

```
    // Decrypt the incoming message
    AES_CFB_decrypt(&dec_key, recv_buf, bytes_rcvd);

    // Parse the message
    err = ParseMessage(recv_buf, bytes_rcvd, output_buffer, sizeof(output_buffer));
    if(err != -1) {
        // Print the string from the target
        output_buffer[16] = 0x0;
        printf("Recevied from target: %s\n", output_buffer);
    }
    else {
        printf("Error in received message, exiting\n");
    }

    close(sock);
    exit(0);
}

// This function simply requests a new AES key from the user, which
// is entered as ASCII text. This isn't the greatest way to create
// a key, but for the purposes of this example, it works just fine.
void GetAESKey(AES_KEY *key) {
    char temp_key[16];

    // Get the AES key
    printf("Enter AES key (must be exactly 16 characters):\n");
    scanf("%16s", temp_key);
    AES_set_encrypt_key(temp_key, 128, &key);
    memset(temp_key, 0x0, sizeof(temp_key));

    // Copy the initialization vector directly into the feedback buffer
    printf("Enter new AES Initialization Vector (must be exactly 16 characters):
\n");
    scanf("%16s", cfb_buf);
}
```

```
// Function to parse an encrypted incoming message and return the message data
int ParseMessage(char * msg, int msg_sz, char * data, int data_sz) {
    int checksum, i;
    float rand_value;
    unsigned int message_counter;
    int data_len;

    // Get the counter from the message
    message_counter =  msg[PKT_CTRLOW_BYTE] +
                    ((int)msg[PKT_CTRLOW_BYTE + 1] << 8) +
                    ((int)msg[PKT_CTRLOW_BYTE + 2] << 16) +
                ((int)msg[PKT_CTRLOW_BYTE + 3] << 24);

    // Compare the message counter to our own counter
    if(message_counter != global_counter) {
        printf("Error: Message counter invalid.\n");
        return -1;
    }

    // Finally, increment the global counter
    ++global_counter;

    // The size of our data is put into the packet (max of 256
    // characters of data)
    data_len = msg[PKT_SIZE_BYTE];

    // Check the length of the received message
    if(data_sz < data_len) {
        printf("Error: Message buffer too small.\n");
        return -1;
    }

    // Ignore the random number and get the data into our buffer
    // Put the data into the message
```

```
    for(i = 0; i < data_len; i++) {

        data[i] = msg[i + PKT_HDR_SIZE];

    }


    // Calculate the checksum byte and compare to the received message
    checksum = 0;
    for(i = 1; i < data_sz + PKT_HDR_SIZE; ++i) {

        checksum += data[i];

    }
    if((char)checksum != msg[PKT_CHKSUM_BYTE]) {

        printf("Error: Message checksum invalid.\n");

        return -1;

    }


    // Return the size of the message
    return data_sz + PKT_HDR_SIZE;

}


// Open SSL does not have a CFB decrypt operation, so roll our own
void AES_CFB_decrypt(AES_KEY *key, char *data, int size)
{

    int i;
    char temp;


    for(i = 0;i < size;i++) {

        if(feedback_index == 0) {

            // CFB decrypt operation uses AESEncrypt
            AES_encrypt(feedback_buf, feedback_buf, key);

        }
        temp = data[i];
        data[i] ^= cfb_buf[cfb_index];
        cfb_buf[cfb_index] = temp;
        cfb_index = (cfb_index + 1) & 0x0f;

    }

}
```

Index